PHILIP'S

STARGAZING
2015

MONTH-BY-MONTH GUIDE
TO THE NORTHERN NIGHT SKY

HEATHER COU

www.philipsastronomy.com
www.philips-maps.co.uk

HEATHER COUPER and NIGEL HENBEST are internationally recognized writers and broadcasters on
astronomy, space and science. They have written more
than 40 books and over 1000 articles, and are the
founders of an independent TV production company
specializing in factual and scientific programming.

Heather is a past President of both the British
Astronomical Association and the Society for Popular
Astronomy. She is a Fellow of the Royal Astronomical
Society, a Fellow of the Institute of Physics and a former
Millennium Commissioner, for which she was awarded
the CBE in 2007. Nigel has been Astronomy Consultant
to *New Scientist* magazine, Editor of the *Journal of the
British Astronomical Association* and Media Consultant to
the Royal Greenwich Observatory.

Heather Couper

Published in Great Britain in 2014 by Philip's,
a division of Octopus Publishing Group Limited
(www.octopusbooks.co.uk)
Endeavour House, 189 Shaftesbury Avenue,
London WC2H 8JY
An Hachette UK Company (www.hachette.co.uk)

TEXT
Heather Couper and Nigel Henbest (pages 6–53)
Robin Scagell (pages 61–64)
Philip's (pages 1–5, 54–60)

Copyright © 2014 Heather Couper and Nigel Henbest
(pages 6–53)

Copyright © 2014 Philip's (pages 1–5, 54–64)

ISBN 978–1–84907–335–6

Nigel Henbest

Title page: *The Rosette Nebula*
(Luke Broom-Lynne/Galaxy)

ACKNOWLEDGEMENTS
All star maps by Wil Tirion/Philip's,
with extra annotation by Philip's.
Artworks © Philip's.

**All photographs courtesy
of Galaxy Picture Library:**
Steve Allan *32;*
Luke Broom-Lynne *1, 8, 28;*
Laurence Dunn *44;*
Stuart Green *36;*
Nick Hart *63;*
James Jefferson *40–41;*
Damian Peach *24;*
Ian Sharp *12;*
Robin Scagell *16, 20, 52, 61, 62, 64;*
Mike Wilson *48.*

CONTENTS

The sight of diamond-bright stars sparkling against a sky of black velvet is one of life's most glorious experiences. No wonder stargazing is so popular. Learning your way around the night sky requires nothing more than patience, a reasonably clear sky and the 12 star charts included in this book.

Stargazing 2015 is a guide to the sky for every month of the year. Complete beginners will use it as an essential night-time companion, while seasoned amateur astronomers will find the updates invaluable.

THE MONTHLY CHARTS

Each pair of monthly charts shows the views of the heavens looking north and south. They are usable throughout most of Europe – between 40 and 60 degrees north. Only the brightest stars are shown (otherwise we would have had to put 3000 stars on each chart, instead of about 200). This means that we plot stars down to third magnitude, with a few fourth-magnitude stars to complete distinctive patterns. We also show the ecliptic, which is the apparent path of the Sun in the sky.

USING THE STAR CHARTS

To use the charts, begin by locating the north Pole Star – Polaris – by using the stars of the Plough (see May). When you are looking at Polaris you are facing north, with west on your left and east on your right. (West and east are reversed on star charts because they show the view looking up into the sky instead of down towards the ground.) The left-hand chart then shows the view you have to the north. Most of the stars you see will be circumpolar, which means that they are visible all year. The other stars rise in the east and set in the west.

Now turn and face the opposite direction, south. This is the view that changes most during the course of the year. Leo, with its prominent 'sickle' formation, is high in the spring skies. Summer is dominated by the bright trio of Vega, Deneb and Altair. Autumn's familiar marker is the Square of Pegasus, while the winter sky is ruled over by the stars of Orion.

The charts show the sky as it appears in the late evening for each month: the exact times are noted in the caption with the chart. If you are observing in the early morning, you will find that the view is different. As a rule of thumb, if you are observing two hours later than the time suggested in the caption, then the following month's map will more accurately represent the stars on view. So, if you wish to observe at midnight in the middle of February, two hours later than the time suggested in the caption, then the stars will appear as they are on March's chart. When using a chart for the 'wrong' month, however, bear in mind that the planets and Moon will not be shown in their correct positions.

THE MOON, PLANETS AND SPECIAL EVENTS

In addition to the stars visible each month, the charts show the positions of any planets on view in the late evening. Other planets may also be visible that month, but they will not be on the chart if they have already set, or if they do not rise until early morning. Their positions are described in the text, so that you can find them if you are observing at other times.

We have also plotted the path of the Moon. Its position is marked at three-day intervals. The dates when it reaches First Quarter, Full Moon, Last Quarter and New Moon are given in the text. If there is a meteor shower in the month, we mark the position from which the meteors appear to emanate – the *radiant*. More information on observing the planets and other Solar System objects is given on pages 54–57.

Once you have identified the constellations and found the planets, you will want to know more about what's on view. Each month, we explain one object, such as a particularly interesting star or galaxy, in detail. We have also chosen a spectacular image for each month and described how it was captured. All of these pictures were taken by amateurs. We list details and dates of special events, such as meteor showers or eclipses, and give observing tips. Finally, each month we pick a topic related to what's on view, ranging from asteroids to white dwarfs and globular clusters, and discuss it in more detail. Where possible, all relevant objects are highlighted on the maps.

FURTHER INFORMATION

The year's star charts form the heart of the book, providing material for many enjoyable observing sessions. For background information turn to pages 54–57, where diagrams help to explain, among other things, the movement of the planets and why we see eclipses.

Although there is plenty to see with the naked eye, many observers use binoculars or telescopes, and some choose to record their observations using cameras, CCDs or webcams. For a round-up of what's new in observing technology, go to pages 61–64, where equipment expert Robin Scagell offers advice on how to cope with light pollution.

If you have already invested in binoculars or a telescope, then you can explore the deep sky – nebulae (starbirth sites), star clusters and galaxies. On pages 58–60 we list recommended deep-sky objects, constellation by constellation. Use the appropriate month's maps to see which constellations are on view, and then choose your targets. The table of 'limiting magnitude' (page 58) will help you to decide if a particular object is visible with your equipment.

Happy stargazing!

If ever there was a time to see A-list stars strutting their stuff, it's January. And this year, the giant planet **Jupiter** joins the dazzling denizens of **Orion, Taurus, Gemini** and **Canis Major** to make up a scintillating celestial tableau.

▼ *The sky at 10 pm in mid-January, with Moon positions at three-day intervals either side of Full Moon. The star positions are also correct for 11 pm at*

JANUARY'S CONSTELLATION

Sparkling overhead on January nights, **Auriga** (the Charioteer) is named after the lame Greek hero Erichthoneus, who invented the four-horse chariot. Its roots date back to the Babylonians, around 1000 BC, who saw Auriga as a shepherd's crook.

Capella, the sixth brightest star in the sky, means 'the little she-goat' – but there's nothing little about this giant yellow star, which is over 150 times more luminous than our Sun, twice as wide, and almost three times heavier. It also holds a substantial yellow companion in thrall.

Nearby, you'll find a tiny triangle of stars nicknamed 'the Kids' (Haedi). Two of the Kids are variable stars: 'eclipsing binaries'. These are stars which change in brightness because a companion star passes in front of them. **Zeta Aurigae** is an orange star eclipsed every 972 days by a blue partner.

Epsilon Aurigae is one of the weirdest star systems known. Every 27 years, it suffers two-year long eclipses, caused by a dark disc of material that's as big as the orbit of Jupiter or Saturn. No two eclipses are the same – and there are tantalizing hints of proto-Jupiter-sized planets within the disc.

Also, bring out those binoculars (better still, a small telescope) to sweep within the 'body' of the Charioteer to find three very pretty open star clusters: **M36, M37** and **M38**.

PLANETS ON VIEW

Beautiful **Venus** has a starring role throughout the first half of 2015, soaring into the sky as the Evening Star, and skimming

the beginning of January, and 9 pm at the end of the month. The planets move slightly relative to the stars during the month.

past five other planets. This month, you'll catch Venus gradually rising up from the sunset twilight, in the south-west at magnitude −3.8.

The first fellow-planet Venus encounters is **Mercury**, putting on its best evening appearance of the year. The innermost planet (magnitude −0.5) lies only a degree to the lower right of Venus for a few days before its greatest eastern elongation on **14 January**.

To the upper left of Venus you'll find **Mars**, setting about 7.45 pm. At magnitude +1.1, the Red Planet tracks from Capricornus to Aquarius during January. Use Mars to locate **Neptune** when it passes very close on **19 January** (see Special Events). The outermost planet lies in Aquarius and sets just before 8 pm.

Uranus, in Pisces, is on the border of naked-eye visibility at magnitude +5.8, and sets around 11.30 pm.

Across the sky from the rest of this planetary activity, **Jupiter** is king of the later evening sky. Rising about 6.30 pm, the giant planet blazes at magnitude −2.4 between the constellations of Cancer and Leo.

Finally, **Saturn** rises at 4 am, at magnitude +0.7 on the borders of Libra and Scorpius.

MOON

The Moon lies near Jupiter on **7 January**, and it's below Regulus on **8 January**. The morning of **13 January** sees the Last Quarter Moon above Spica. Before dawn on **16 January**, the crescent

WEST

Uranus
PISCES
TRIANGULUM
PERSEUS
ARIES
Mira
CETUS
Pleiades
26 Jan
Aldebaran
TAURUS
ERIDANUS
epsilon
zeta
2 Jan
Rigel
LEPUS
Capella
M38
M36
Betelgeuse
ORION
Mirzam
COLUMBA
Zenith
AURIGA
M37
GEMINI
Rosette Nebula
CANIS MAJOR
Adhara
SOUTH
Castor
Pollux
5 Jan
Procyon
THE MILKY WAY
Sirius
URSA MAJOR
CANCER
CANIS MINOR
PUPPIS
The Sickle
Jupiter
Regulus
Ecliptic
8 Jan
LEO
HYDRA
VIRGO
EAST
SE
MS

January's Picture
Rosette Nebula

Radiant of
Quadrantids

	MOON		
	Date	Time	Phase
Jupiter	5	4.53 am	Full Moon
Uranus	13	9.46 am	Last Quarter
	20	1.14 pm	New Moon
Moon	27	4.48 am	First Quarter

Moon lies near Saturn. On **22 January**, you'll find the slim crescent Moon above Venus, with Mars to its upper left. On **29 January**, the Moon is close to Aldebaran.

SPECIAL EVENTS

On **4 January**, at 6.36 am, the Earth is at perihelion, its closest point to the Sun – a 'mere' 147 million kilometres away.

The night of **3/4 January** sees the maximum of the **Quadrantid** meteor shower, tiny particles of dust shed by the old comet 2003 EH$_1$ that burn up – often in a blue or yellow streak – as they enter the Earth's atmosphere. Unfortunately, all but the brightest meteors will be drowned out by moonlight this year.

On **19 January**, Neptune lies only 20 arc minutes above Mars. You'll need optical aid to see this faint world (magnitude +7.9); a good telescope shows the contrast between the two most strongly coloured planets – bluish Neptune and the famous Red Planet.

JANUARY'S OBJECT

This month is your chance to spot elusive Mercury, which never strays far from the Sun's glare, and so upstage the architect of our Solar System, Nicolaus Copernicus, who was thwarted from observing the planet by mists from the River Vistula.

Only one-third of Earth's distance from the Sun, Mercury's surface is baked to 450°C at noon, dropping to −180°C at night. A 'day' – from noon to noon – on this slowly rotating world lasts 176 days, twice as long as its 'year' of 88 Earth-days.

The pioneering space probe Mariner 10 sent back brief images as it swung past Mercury in 1974. It imaged a wrinkled, gnarled surface – the result of Mercury's surface crumpling up as it has cooled and shrunk (like a dried-up apple). Mariner revealed that this mini-world – just one-third the Earth's diameter – is peppered in craters, a legacy from an intense cosmic bombardment in its youth.

▶ *The Rosette Nebula, captured by Luke Broom-Lynne in Attleborough, Norfolk. He used a Canon 450DSLR camera attached to a 250 mm f/4.7 Newtonian telescope.*

But the floodgates opened in 2011, when NASA's Messenger probe went into orbit around Mercury. Named for Mercury's mythological role as fleet-footed messenger to the gods (as well as a convoluted acronym), Messenger has seen evidence for past volcanic activity, plus water in Mercury's thin atmosphere. Most excitingly, peering inside permanently shaded craters at the planet's north pole, Messenger has found signs of organic compounds and ice sheets up to 20 metres thick.

JANUARY'S PICTURE

A stunning sight in the faint constellation Monoceros, the aptly named **Rosette Nebula** is a hotbed of star formation, lying 5000 light years away. This beautiful close-up reveals the central cluster of young stars, whose searing radiation is energizing their natal gas. Some 130 light years in diameter, the Rosette may weigh as much as 10,000 Suns.

JANUARY'S TOPIC
White dwarfs

Brilliant **Sirius** – sailing high in our southern skies this month – is accompanied by a most extraordinary companion star. Discovered by telescope-maker Alvan Clark in 1862, its existence had been predicted by Friedrich Bessel almost 20 years earlier, when he noticed that the Dog Star was being tugged out of position by a then-unseen body. This small companion to our celestial canine is fondly nicknamed 'the Pup'.

The Pup isn't a particularly faint star – it shines at magnitude +8.5, so you should be able to spot it in binoculars. But Sirius glares at magnitude −1.47, and the contrast makes the Pup a devil to find: you'll need a telescope with a mirror at least 300 mm across.

The Pup is only the size of the Earth, yet it boasts 98% the mass of our Sun. Hence it's called a 'white dwarf' – a species that may make up 10% of all the stars in the Galaxy.

A white dwarf is the core of an old star that has died, having puffed off its atmosphere as a brief – but glorious – planetary nebula. No longer supported by nuclear reactions, the core has collapsed under its own gravity, squeezing electrons and the nuclei of atoms to unbelievable densities. A matchbox full of white-dwarf material would outweigh an elephant!

With no power source, a white dwarf will eventually just fade away, to end up as a cold, black cinder. Seven billion years hence, this is the fate that awaits our Sun.

The first signs of spring are now on the way. The winter star patterns are drifting towards the west, as a result of our annual orbit around the Sun. Imagine: you're on a fairground carousel, circling round on your horse, and looking out around you. At times you spot the ghost train; sometimes you see the roller-coaster; and then you swing past the candy-floss stall. So it is with the sky: as we circle our local star, we get to see different stars and constellations with the changing seasons.

▼ The sky at 10 pm in mid-February, with Moon positions at three-day intervals either side of Full Moon. The star positions are also correct for

FEBRUARY'S CONSTELLATION

You can't ignore **Gemini** in February. High in the south, the constellation is crowned by the stars **Castor** and **Pollux**, similar in brightness and representing the heads of a pair of twins – their stellar bodies running in parallel lines of stars towards the west. Legend has it that Castor and Pollux were twins. On the night Princess Leda married the king of Sparta, wicked old Zeus (Jupiter) invaded the marital suite, disguised as a swan. Pollux was Jupiter's son – and was therefore immortal – while Castor was merely human. But the pair were so devoted to each other that Zeus placed both Castor and Pollux amongst the stars.

Castor is an amazing star: a family of six. Even through a small telescope, you can see that Castor is a double star. Both of these stars are themselves double (although you need special equipment to detect this). Then there's another outlying star, visible through a telescope, which also turns out to be double.

PLANETS ON VIEW

Jupiter lives up to its title of King of the Planets, resplendent all night long between Cancer and Leo. At magnitude −2.3, the giant world (see this month's Object) dominates the night sky as it comes closest to the Earth this year, reaching opposition on **6 February**.

11 pm at the beginning of February, and 9 pm at the end of the month. The planets move slightly relative to the stars during the month.

Even more brilliant **Venus** (magnitude −3.8) is now steaming upwards in the evening twilight – by the end of the month, the Evening Star is setting almost three hours after the Sun. And the Goddess of Love is continuing her liaisons with other planets….

Venus sails past **Neptune** – with less than a degree to spare – on **1 February** (see Special Events). The most distant planet lies in Aquarius and sets around 6 pm.

And the Evening Star is also chasing **Mars**, as the Red Planet moves from Aquarius to Pisces, setting at 8 pm. They have a brief encounter on **21 and 22 February** (see Special Events).

Uranus (magnitude +5.9) also lies in Pisces, slipping below the horizon around 9.30 pm.

Early birds can catch **Saturn** (magnitude +0.7), in Scorpius, rising at 2 am in the south-east.

Mercury reaches greatest western elongation on **24 February**, but – as seen from Britain – it's lost in the morning twilight glow.

MOON

The brilliant object lying near the Full Moon on **3 February** is Jupiter. The waning Moon lies near Spica on the night of **9/10 February**. Saturn lies just to the right of the Moon in the morning skies of **13 February**. On **20 and 21 February** you may catch the slenderest crescent Moon after sunset, near Venus and Mars. The night of **25/26 February** sees the Moon just graze past Aldebaran (see Special Events).

MOON		
Date	Time	Phase
3	11.09 pm	Full Moon
12	3.50 am	Last Quarter
18	11.47 pm	New Moon
25	5.14 pm	First Quarter

February's Object
Jupiter

Jupiter

Moon

WEST

EAST

SOUTH

PISCES
CETUS
PERSEUS
TAURUS
Pleiades
Aldebaran
ERIDANUS
LEPUS
ORION
Rigel
CANIS MAJOR
Adhara
Betelgeuse
Sirius
AURIGA
Capella
Zenith
GEMINI
Castor
Pollux
Procyon
CANIS MINOR
THE MILKY WAY
PUPPIS
URSA MAJOR
Jupiter
The Sickle
LEO
Regulus
HYDRA
VIRGO
Ecliptic

24 Feb
27 Feb
3 Feb
6 Feb

11

On **1 February**, Venus passes less than a degree from Neptune, providing a rare chance to compare directly the brightest and faintest planets. Through good binoculars or a telescope, you can spot Neptune as the fainter of the two 'stars' to the upper right of Venus. At magnitude +8.0, Neptune is 50,000 times fainter than Venus.

Venus has a tryst with Mars, just a week too late for Valentine's Day, on **21 and 22 February**. As these worlds pass – just half a degree apart – the Red Planet (magnitude +1.2) is outshone a hundred times over by gaudy Venus.

On the night of **25/26 February**, the Moon passes very close to Aldebaran; as seen from Orkney and Shetland, the star disappears behind the Moon at 0.15 am.

▲ Ian Sharp, observing at Ham in West Sussex, used a 125 mm f/7 APO refractor with a QHY5L-II camera operating at 30 frames per second to capture this sequence of images of the ISS.

FEBRUARY'S OBJECT

Jupiter reaches opposition on **6 February**, lying opposite the Sun, and is at its closest to the Earth. 'Close' is a relative term, though! The giant planet lies about 700 million kilometres away.

At 143,000 kilometres in diameter, Jupiter could contain 1300 Earths – and the cloudy gas giant is very efficient at reflecting sunlight. Jupiter is shining at a dazzling magnitude −2.3, so bright it can cast a shadow, and is a fantastic target for stargazers, whether you're using your unaided eyes, binoculars or a small telescope.

Despite its size, Jupiter spins faster than any other planet, in 9 hours 55 minutes. As a result, its equator bulges outwards – through a small telescope, it looks like a tangerine crossed with an old-fashioned humbug. The stripes are cloud belts of ammonia and methane stretched out by the planet's dizzy spin.

◉ **Viewing tip**

Don't think that you need a telescope to bring the heavens closer. Binoculars are excellent – and you can fling them into the back of the car at the last minute. But when you buy binoculars, make sure that you get those with the biggest lenses, coupled with a modest magnification. Binoculars are described, for instance, as being '7×50' – meaning that the magnification is seven times, and that the diameter of the lenses is 50 millimetres. These are ideal for astronomy – they have good light grasp, and the low magnification means that they don't exaggerate the wobbles of your arms too much. It's always best to rest your binoculars on a wall or a fence to steady the image. Some amateurs are the lucky owners of huge binoculars – say, 20×70 – with which you can see the rings of Saturn (being so large, these binoculars need a special mounting). But above all, *never* buy binoculars with small lenses that promise huge magnifications – they're a total waste of money.

Jupiter commands a family of almost 70 moons. The four biggest are visible in good binoculars, and even – to the really sharp-sighted – to the unaided eye. These are worlds in their own right – Ganymede is even bigger than the planet Mercury. But two vie for 'star' status: Io and Europa. Io is erupting, with incredible volcanoes erupting plumes of sulphur dioxide 300 kilometres into space. Beneath a solid ice surface, brilliant white Europa probably contains oceans of liquid water, where alien fish may swim....

FEBRUARY'S PICTURE

In this stunning sequence of images, Ian Sharp has captured the International Space Station passing right in front of the Moon, on 9 April 2014.

Though this event is rare, it's quite common to see the ISS crossing the sky as a brilliant point. There's an excellent opportunity during the first half of this month: about 6.10 pm every evening, the ISS rises in the west, to pass almost over-head a few minutes later. At magnitude −4.4, the ISS outshines everything else in the sky, including Venus.

FEBRUARY'S TOPIC
Asteroids

In 1801, observing from Sicily, Guiseppe Piazzi stumbled across an object circling the Sun between Mars and Jupiter. His 'new planet' Ceres was soon joined by other diminutive worlds, and William Herschel (the discoverer of Uranus) dubbed them 'asteroids' – meaning 'star-like'.

Numbering in their millions, the asteroids are irregular chunks of rock and iron that failed to assemble into a planet because of Jupiter's mighty gravity. At 975 kilometres across, Ceres is by far the largest of these worldlets – it makes up one-third of the mass of the asteroid belt.

Asteroids may also have been crucial to the origin of life. Billions of years ago, they delivered water, and possibly the building blocks of living cells, to the Earth.

So far, space probes have visited ten asteroids. NASA's DAWN craft circled Vesta for a year (at magnitude +5.1, Vesta is the brightest asteroid, and the only one visible to the unaided eye). DAWN discovered a heavily cratered mini-world, with gullies possibly cut by running water.

This month, DAWN begins to home in on Ceres, and will slip into orbit in March or April. From an altitude of 5900 kilometres, its height will reduce to a mere 700 kilometres within a year. Expect some incredible close-up images, and amazing data!

We're treated to a major partial eclipse of the Sun this month: wherever are you in the British Isles, you'll see over 80% of the Sun's face covered by the Moon – rising to 98% if you're as far north as Shetland. Travel to the Faroes or Svalbard to experience the magic of a total eclipse!

▼ *The sky at 10 pm in mid-March, with Moon positions at three-day intervals either side of Full Moon. The star positions are also correct for 11 pm at*

MARCH'S CONSTELLATION

Cancer is hardly one of the most spectacular constellations in the sky. Although it lies in the Zodiac – the band through which the Sun, Moon and planets appear to move – its stars are so faint that city lights completely drown them out. Under dark skies, look between '**the Sickle**' of **Leo** and the twin stars **Castor** and **Pollux** in **Gemini,** and you'll locate the slender little constellation.

In legend, Cancer is the crab that attempted to nip Hercules during his altercation with the multi-headed monster Hydra – one of his '12 labours' ordered by the Oracle at Delphi. Alas – Hercules crushed the crustacean under his foot. But Juno (Jupiter's wife) took pity on the crab and placed it in the sky.

At the centre of the constellation is a gem – the aptly titled '**Beehive Cluster**'. Officially known as Praesepe, it's a swarm of perhaps 1000 stars lying nearly 600 light years away. The Beehive is easily visible to the unaided eye, and was well known to ancient Greek astronomers such as Aratos, Hipparchus and Ptolemy.

PLANETS ON VIEW

Brilliant **Venus** rules the evening sky in the west, blazing at magnitude −3.9. The Evening Star slips below the horizon at 8.30 pm at the start of March, and as late as 11 pm after the clocks change at the end of the month.

At the beginning of March, you'll find **Mars** (magnitude +1.3) below Venus: the Red Planet lies in Pisces and sets just after 8 pm.

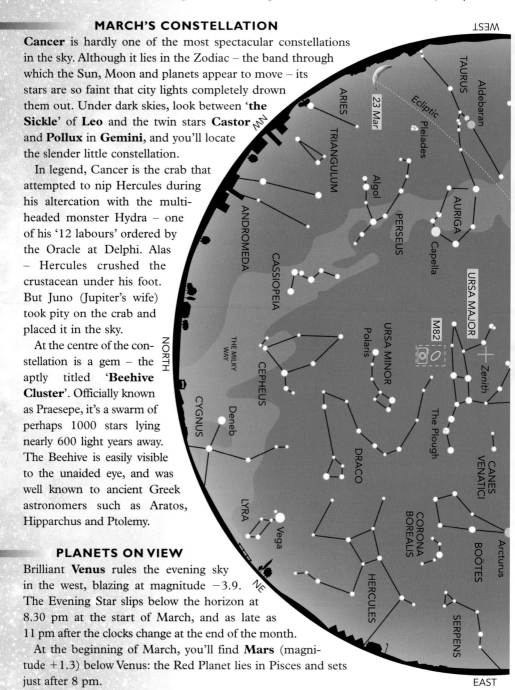

the beginning of March, and 10 pm at the end of the month (after BST begins). The planets move slightly relative to the stars during the month.

Faint **Uranus** (magnitude +5.9 in Pisces) is having some adventures this month – with Venus on **4 March** and Mars on **11 March** (see Special Events). It's so low in the west, setting around 8 pm, that you'll need a small telescope to follow its antics.

Jupiter is visible all night long and outshines all the stars at a magnificent magnitude −2.3. The giant planet is currently inhabiting the faint constellation of Cancer. Use binoculars, held steadily, to observe its four biggest moons.

Wait up till 1 am to see the final planet of the night, **Saturn**, rising in the south-east. It shines at magnitude +0.5 in Scorpius.

Mercury and **Neptune** are lost in the Sun's glare this month.

MOON

The Moon lies near Jupiter on **2 and 3 March**, and passes below Regulus on **4 March**. The star below the Moon on **8 March** is Spica. You'll find the Moon near to Saturn on the morning of **12 March**. On **22 March**, the thin crescent Moon lies to the lower left of Venus. The Moon is keeping company with Jupiter again on **29 and 30 March**, and Regulus on **31 March**.

SPECIAL EVENTS

Venus has an incredibly close encounter with Uranus on **4 March**, passing just 6 arc minutes away. You'll find Uranus (use a small telescope) to the lower left of the brilliant Evening Star, and almost 10,000 times fainter.

WEST

ERIDANUS
Aldebaran
TAURUS
Rigel
ORION
Betelgeuse
26 Mar
GEMINI
AURIGA
Castor
Pollux
Beehive Cluster
CANCER
Jupiter
2 Mar
Zenith
URSA MAJOR
The Sickle
Regulus
LEO
CANES VENATICI
Denebola
5 Mar
VIRGO
8 Mar
BOÖTES
Arcturus
Spica
Ecliptic
SERPENS

LEPUS
Sirius
Procyon
CANIS MINOR
THE MILKY WAY
CANIS MAJOR
PUPPIS
HYDRA
CORVUS
SOUTH
SE

EAST

March's Object
Regulus

March's Picture
M82

Jupiter

Moon

MOON		
Date	**Time**	**Phase**
5	6.05 pm	Full Moon
13	5.48 pm	Last Quarter
20	9.36 am	New Moon
27	7.42 am	First Quarter

It's Mars's turn to brush past Uranus on **11 March**. Again, Uranus lies to the lower left of the brighter planet, at a distance of 17 arc minutes; it's 70 times fainter than the Red Planet.

On **20 March**, a total eclipse of the Sun is visible from parts of the Arctic Ocean, including the Faroe Islands and Svalbard. It appears as a partial eclipse from north-west Asia, northern Africa and all of Europe, including the British Isles (see this month's Topic).

The Vernal Equinox, on **20 March** at 10.45 pm, marks the beginning of spring, as the Sun moves up to shine over the northern hemisphere.

29 March, 1.00 am: British Summer Time starts – don't forget to put your clocks forward (the mnemonic is 'Spring forward, Fall back').

This month, the DAWN spacecraft goes into orbit around Ceres, the largest object in the asteroid belt (see February's Topic).

MARCH'S OBJECT

Regulus – the 'heart' of Leo (the Lion) – appears to be a bright but fairly anonymous star. Some 77 light years away, it's young (a few hundred million years old), 3.8 times heavier than the Sun, and it chucks out 300 times as much energy as our local star. But recent discoveries have revealed it to be a maverick. It spins in less than a day – meaning that it has a rotational velocity of over 1 million km/h. If it were to spin only 10% faster than this, it would tear itself apart. This bizarre behaviour means that its equator bulges like a tangerine (its equatorial girth is one-third larger than its north–south diameter). To

▼ *Robin Scagell captured this image of M82 with an 80 mm refractor from Flackwell Heath, Buckinghamshire, using an Atik 314L+ CCD camera through RGB filters. The total exposure time was 45 minutes.*

◉ **Viewing tip**
This is the time of year
to tie down your compass
points – the directions
of north, south, east and
west as seen from your
observing site. North is
easy – just latch on to
Polaris, the Pole Star
(see May's Constellation).
And at noon, the Sun is
always in the south. But
the useful extra in March
is that we hit the Spring
(Vernal) Equinox, when
the Sun rises due east,
and sets due west. So
remember those positions
relative to a tree or house
around your horizon.

compound it all, its rotation axis is tilted at an angle of 86 degrees – which means that Regulus zaps through the Milky Way virtually on its side!

MARCH'S PICTURE

Some 12 million light years away, baby-boomer galaxy **M82** is a unique sight. Its ragged appearance has long defied description, but now we know that it is a spiral galaxy stirred up by the gravity of its companion, the serene spiral M81. The pair are a glorious sight through a wide-field telescope in **Ursa Major** (see also May's Object). Violent star formation has its downside. Massive young stars live fast and furiously, exploding as supernovae. The object near M82's centre (arrowed) is a star that was seen to explode on 21 January 2014. It was the closest supernova in ten years.

MARCH'S TOPIC
Solar eclipse

At 8.39 am on **20 March**, the Faroe Islands will be plunged into darkness. The culprit is a total eclipse of the Sun – the closest such eclipse to British soil until September 2090. Although the Faroe Islands aren't British (they're Danish), they are within easy travelling distance of the mainland – so expect a party!

Total solar eclipses take place when the Moon completely overlaps the Sun in the sky. The coincidence is that the Sun is 400 times wider than the Moon – but it is 400 times further away.

Throughout Britain, most of the Sun will be covered up. Even as far south as Kent, 86% of the Sun disappears behind the Moon – while in the Outer Hebrides, it's 98%. Don't expect to see the Sun's delicate outer atmosphere during this partial eclipse: there's still too much sunlight around. Instead, expect to see the sky darken markedly, and be prepared to experience dampness and chill.

The Faroe Islands get the full monty. If it's clear – a big 'if', because weather conditions in March can be pretty appalling – you'll see the Sun transformed into a Chinese dragon-mask, the tendrils of its pearly atmosphere wrapped around the black body of the Moon. And you may spot the planets Venus, Mars and Mercury, along with the three stars making up the Summer Triangle.

Totality lasts just 2 minutes 2 seconds – then it's all over. Even if you are clouded out, don't despair: at this latitude you're in with a good chance of seeing the Northern Lights. One way or another, you should get a light show!

We have a celestial jewel this month – the glorious Evening Star. Brighter than anything in the night sky bar the Moon, **Venus** hangs like a tiny lantern in the west, staying above the horizon until after midnight. Its slightly fainter rival, **Jupiter**, lies near the ancient constellation of **Leo**: the shape of the celestial lion does resemble its namesake, which is more than can be said of its fellow spring constellation, **Virgo** (the Virgin). It's hard to imagine her as anything but a giant letter 'Y'!

APRIL'S CONSTELLATION

The 'Y'-shaped constellation of **Virgo** is the second largest in the sky. It takes a bit of imagination to see the group of stars as a virtuous maiden holding an ear of corn (the bright star **Spica**), but this very old constellation has associations with the times of harvest. In the early months of autumn, the Sun passes through the stars of Virgo, hence the connections with the gathering-in of fruit and wheat.

Spica is a hot, blue-white star over 12,000 times brighter than the Sun, boasting a temp-erature of 22,500°C. It has a stellar companion, which lies just 18 million kilometres away from Spica – closer than Mercury orbits the Sun. Both stars inflict a mighty gravitational toll on each other, raising enormous tides and thus creating two distorted, egg-shaped stars. In fact, Spica is the celestial equivalent of a rugby ball.

The glory of Virgo is the 'bowl' of the 'Y' shape. Scan the upper region with a small telescope – at a low magnification – and you'll find it packed with faint, fuzzy blobs. These are just a few of the myriad galaxies – star-cities like the Milky Way – that make up the gigantic **Virgo Cluster** (see this month's Topic).

▼ The sky at 11 pm in mid-April, with Moon positions at three-day intervals either side of Full Moon. The star positions are also correct for midnight at the beginning of

WEST

TAURUS
ORION
Betelgeuse
Pleiades
Venus
Ecliptic
22 Apr
GEMINI
AURIGA
Castor
Pollux
Algol
Capella
PERSEUS
URSA MAJOR
ANDROMEDA
CASSIOPEIA
URSA MINOR
Polaris
Kochab
The Plough
Zenith
NORTH
CEPHEUS
DRACO
BOÖTES
CORONA BOREALIS
THE MILKY WAY
Deneb
CYGNUS
Vega
Radiant of Lyrids
LYRA
HERCULES
OPHIUCHUS

EAST

April, and 10 pm at the end of the month. The planets move slightly relative to the stars during the month.

WEST

THE MILKY WAY

GEMINI

Procyon

CANIS MINOR

Castor

Pollux

25 Apr

CANCER

Jupiter

URSA MAJOR

The Sickle

Regulus

HYDRA

LEO

28 Apr

Denebola

CORVUS

Zenith

CANES VENATICI

M87

Arcturus

Virgo Cluster

4 Apr

The Plough

CORONA BOREALIS

BOÖTES

SERPENS

VIRGO

Spica

Ecliptic

LIBRA

HERCULES

OPHIUCHUS

SE

EAST

SW

SOUTH

PLANETS ON VIEW

The two most brilliant planets are vying for our attention this month. In the west, **Venus** is speeding upwards through Taurus, passing the Pleiades (Seven Sisters) on **11 April**. The Evening Star grows steadily brighter to reach magnitude −4.0 by the end of April, when it's setting after midnight. During the evening – when Venus is visible in a perfectly dark sky – check if you can see the brilliant planet casting shadows.

In the south, **Jupiter** is putting on a rival show: though it's fainter than Venus, at magnitude −2.1, it's much higher in the sky. Lying in Cancer, the giant planet is visible right through to the morning twilight.

At the start of April, you may catch **Mars** in the twilight, well below Venus: the Red Planet is slipping downwards, to disappear by the end of the month.

Mars passes **Mercury** – travelling upwards – on **22 April**, when Mars is at magnitude +1.4, well outclassed by Mercury at magnitude −1.1. As it climbs steadily higher, Mercury becomes easier to spot, appearing right next to the Pleiades on **30 April**.

Saturn, in Scorpius, is rising about 11.30 pm. At magnitude +0.4, the pale yellow ringworld is appreciably brighter than the constellation's main star (which lies to its lower left) – the red giant Antares.

Uranus and **Neptune** are lost in the Sun's glare during April.

April's Object and Picture Venus

Radiant of Lyrids

Venus

Jupiter

Moon

MOON		
Date	Time	Phase
4	1.05 pm	Full Moon
12	4.44 am	Last Quarter
18	7.57 pm	New Moon
26	0.55 am	First Quarter

MOON

The Full Moon lies above Spica on **4 April**. The bright object near the Moon on the mornings of **8 and 9 April** is Saturn. On **21 April**, the crescent Moon lies 7 degrees to the lower left of Venus, and on **22 April** it's well to the left of the Evening Star. The First Quarter Moon passes below Jupiter on **26 April**, and Regulus on **27 April**.

SPECIAL EVENTS

People in eastern Asia, Australasia, the Pacific Ocean and western parts of North America are treated to a total eclipse of the Moon on **4 April**. It's not visible from Britain.

22/23 April: It's the maximum of the **Lyrid** meteor shower, which – by perspective – appears to emanate from the constellation of Lyra. The shooting stars are dust particles from Comet Thatcher. The best views will be after midnight, when the Moon has set.

APRIL'S OBJECT

Venus – the planet of love – is resplendent in our evening skies this month (see 'Viewing tip'). So brilliant and beautiful, she can even cast a shadow in a really dark, transparent sky.

▲ *The Moon and Venus in the evening sky, photographed by Robin Scagell in Flackwell Heath, Buckinghamshire. He used a Canon 10D DSLR, with an 8-second exposure. This image was captured on 23 April 2004, and is very similar to the appearance that the pair will make on 22 April 2015.*

Her purity and lantern-like luminosity are beguiling – but looks are deceptive. Earth's twin in size, Venus could hardly be more different from our warm, wet world. The reason for the planet's brilliance is the highly reflective clouds that cloak its surface: probe under these palls of sulphuric acid hanging in an atmosphere of carbon dioxide, and you find a planet out of hell. Volcanoes are to blame. They have created a runaway 'greenhouse effect' that has made Venus the hottest and most poisonous planet in the Solar System. At 460°C, this world is hotter than an oven. The pressure at its surface is around 90 Earth-atmospheres. So – if you visited Venus – you'd be simultaneously roasted, crushed, corroded and suffocated!

APRIL'S PICTURE

The planet **Venus**, resplendent this month, is the brightest object in the night sky after the Moon. This lovely shot captures the pair in a twilight countryside sky. Like the Moon, Venus shows phases – from crescent to full – as it circles the Sun. You'll need a telescope to spot these, but Venus is a glorious sight to the unaided eye: you can even see the planet in daylight.

APRIL'S TOPIC
Clusters of galaxies

The fuzzy blobs that a backyard telescope reveals in the 'bowl' of Virgo are just a handful of the thousands of galaxies making up the **Virgo Cluster** – our closest giant cluster of galaxies, lying at a distance of almost 55 million light years.

Galaxies are gregarious. Thanks to gravity, they like living in groups. Our Milky Way and the neighbouring giant spiral, the Andromeda Galaxy, are members of a small cluster of about 30 smallish galaxies called the Local Group.

But the Virgo Cluster is in a different league: it's like a vast galactic swarm of bees. What's more, its enormous gravity holds sway over the smaller groups around – including our Local Group – making up a cluster of clusters of galaxies, the Virgo Supercluster.

The 2000 galaxies in the Virgo Cluster are also mega. Many are spirals like our Milky Way, but some are even more spectacular. The heavyweight of the cluster is **M87**, a giant elliptical galaxy emitting a jet of gas over 5000 light years long that travels at one-tenth the speed of light.

Galaxy clusters and superclusters are the salt of the Universe. They make up the structure of the cosmos, with huge filaments of superclusters enclosing enormous empty voids – like a gigantic Swiss cheese.

A sure sign that warmer days are here is the appearance of **Arcturus** – a distinctly orange-coloured star that lords it over a huge area of sky devoid of other bright stars. It's the brightest star in the constellation of **Boötes** (the Herdsman), who shepherds the two bears – **Ursa Major** and **Ursa Minor** – through the heavens. Summer is on the way!

▼ *The sky at 11 pm in mid-May, with Moon positions at three-day intervals either side of Full Moon. The star positions are also correct for midnight at the beginning of*

MAY'S CONSTELLATION

Ursa Major (the Great Bear) is an internationally favourite star pattern. In Britain, its seven brightest stars are called 'the Plough'. Most people today have never seen an old-fashioned horse-drawn plough, though, and we've found children naming this star pattern 'the saucepan.' In North America, it's known as 'the Big Dipper'.

Even so, the Plough is the first constellation that most people get to know. There are two reasons. First, it is always on view in the northern hemisphere. And second, the two end stars of the 'bowl' of the Plough point directly towards the Pole Star, **Polaris**.

Ursa Major is unusual in a couple of ways. It contains a double star that you can actually split with the naked eye. **Mizar**, the star in the middle of the bear's tail (or the handle of the saucepan) has a fainter companion, **Alcor**. The whole system – once thought to be a chance alignment – consists of six stars.

And – unlike most constellations – the majority of the stars in the Plough lie at the same distance and were born together. Leaving aside the two end stars, **Dubhe** and **Alkaid**, the others are all moving in the same direction (along with brilliant Sirius, which is also a member of the group). Over thousands of years, the shape of the Plough will gradually change, as Dubhe and Alkaid go off on their own paths.

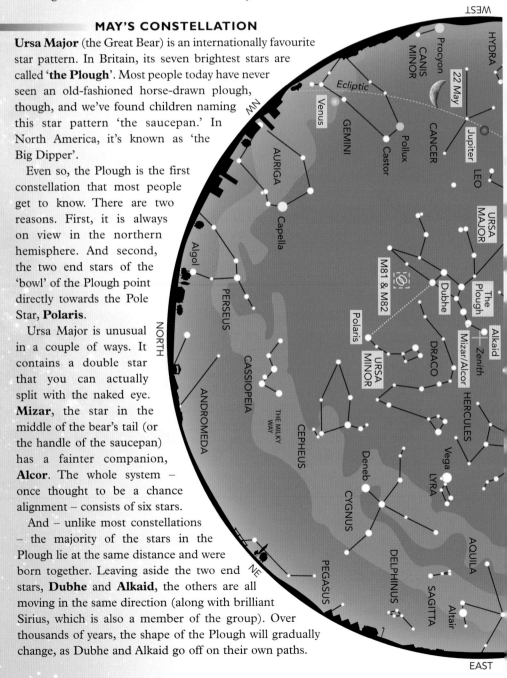

May, and 10 pm at the end of the month. The planets move slightly relative to the stars during the month.

PLANETS ON VIEW

Brilliant **Venus** hangs like a glowing lantern in the western sky after sunset. Setting at 0.30 am, the Evening Star is dazzling at magnitude −4.1. During the month, Venus travels through Taurus and Gemini, passing close to the twin stars Castor and Pollux on the last couple of nights of May.

This trajectory is bringing Venus ever closer to **Jupiter**, for a cosmic encounter next month. The giant planet – at magnitude −1.9, second only to Venus in brilliance – lies in Cancer, and is setting around 2 am.

During the first half of May, look low down on the north-western horizon at about 10 pm to spot tiny **Mercury**. On **1 May**, it's just to the left of the Pleiades. As Mercury rises into darker skies – reaching greatest eastern elongation on **7 May** – it fades precipitously, from magnitude −0.3 to +0.4.

These planetary events are overshadowing poor old **Saturn**, which ought to be headlining this month, as it's at opposition on **23 May**. But the ringworld is low in the sky, on the border of Scorpius and Libra, and it shines at a relatively feeble magnitude +0.2. Saturn is visible throughout these short May nights, and a small telescope reveals its magnificent rings and the biggest moon, Titan.

Neptune (magnitude +7.9) is low in the south-eastern sky before dawn. Rising at 3 am, the most distant planet lies in Aquarius. **Mars** and **Uranus** are too close to the Sun to be visible this month.

May's Objects
M81 and M82

May's Picture
Saturn

	MOON		
	Date	Time	Phase
Venus	4	4.42 am	Full Moon
Jupiter	11	11.36 am	Last Quarter
Saturn	18	5.13 am	New Moon
Moon	25	6.19 pm	First Quarter

MOON

The Moon is near Spica on **2 May**. The planet Saturn lies to the right of the Moon on **5 May**. The thin crescent Moon forms a lovely sight with Venus on the evening of **21 May**; it lies below Jupiter on **23 May**. The Moon returns to Spica on **29 May**.

SPECIAL EVENTS

Shooting stars from the Eta Aquarid meteor shower – tiny pieces shed by Halley's Comet and burning up in Earth's atmosphere – will fly across the sky on the nights of both **5/6 and 6/7 May**. Unfortunately, this year the display is spoilt by moonlight.

MAY'S OBJECT

A pair of galaxies this month, nuzzling up close and personal to each other in **Ursa Major**: **M81** and **M82** are visible through good binoculars on a really dark night, though you'll need a moderately powerful telescope to reveal them in detail. 'M' stands for Charles Messier, an 18th-century Parisian astronomer who catalogued 103 'fuzzy objects' that misleadingly resembled comets. He did find a handful of comets – but today, he's better remembered for his *Messier Catalogue*.

M81 is a beautiful, smooth spiral galaxy with curving spiral arms wrapped around a softly glowing core. A twin to our own Galaxy, M81 lies nearly 12 million light years away.

Lying close by is M82 – a spiral galaxy that couldn't be more different. It looks a total mess, with a huge eruption taking place at its core. Some 300 million years ago, a close encounter with its bigger sibling, M81, ripped out streams of gas that are

⊙ *Viewing tip*

It's best to view your favourite objects when they're well clear of the horizon. When you look low down, you're seeing through a large thickness of the atmosphere – which is always shifting and turbulent. It's like trying to observe the outside world from the bottom of a swimming pool! This turbulence makes the stars appear to twinkle. Low-down planets also twinkle – although to a lesser extent, because they subtend tiny discs, and aren't so affected.

still raining back down on to M82's core, creating an explosion of star formation – making M82 the prototype 'starburst galaxy'.

In January 2014, a supernova exploded in M82 (see March's Picture) – the closest to Earth since 2004. Reaching magnitude +10.6, the supernova was visible through a small telescope. It was the result of a star dumping excessive gas on to a white dwarf companion, so tipping its mass over the limit and detonating the tiny star.

MAY'S PICTURE

Glorious ringworld **Saturn** takes centre-stage in the heavens this month, reaching opposition on **23 May**, when it's due south at midnight. You'll need a small telescope to see the rings, which are made of billions of icy chunks. The dramatic dark gap in Saturn's rings visible in this image is the Cassini Division – a region swept clear by the gravity of Saturn's moons. If you have a medium-sized telescope, monitor the planet's gas clouds for signs of activity – it seems to be livening up!

MAY'S TOPIC
Afterglow of the Big Bang

Fifty years ago this month, astronomers published their findings on a discovery that changed forever our understanding of how the Universe began. But it didn't get off to an auspicious start. 'We frankly did not know what to do with our records,' despaired Arno Penzias, 'knowing, at the time, that no astronomical explanation was possible.'

Penzias and his colleague Robert Wilson were surveying gas in the outer Milky Way with a sensitive 'low-noise' radio telescope in Holmdel, New Jersey. Except that there was a noise – a persistent one. The researchers put it down to pigeon-poo accumulating in their giant horn-shaped instrument – and discouraged the offending birds 'by decisive means'.

Still the noise continued. It was spread out evenly all over the sky – and corresponded to a temperature of 2.7° above absolute zero. Unknown to Penzias and Wilson, though, a team of theoreticians at Princeton University, led by Bob Dicke and Jim Peebles, had calculated that – if the cosmos had been born in a mighty fireball – a glimmer of its ferocious heat would linger today, cooled to virtually nothing by the expansion of the Universe.

And – fortuitously – that's what Penzias and Wilson had discovered: the afterglow of creation. It was proof that the Universe was born in a Big Bang, where the pressures and temperatures were almost infinite. And – 13.8 billion years on – its echoes are with us even today.

◀ *Saturn, 15 March 2006: Damian Peach used his 355 mm Schmidt-Cassegrain (C14) telescope with a Lumenera webcam and RGB filters to capture this image. He later processed using Registax 4 software – a stacking programme. The location was Loudwater, Buckinghamshire.*

There's a magnificent planetary spectacle in store later this month, low down in the north-west, as the two most brilliant planets – **Venus** and **Jupiter** – pass so close they almost seem to touch. But this isn't the best month for stargazing. The Sun reaches its highest position over the northern hemisphere in June, giving us the longest days and the shortest nights. But take advantage of the soft, warm weather to acquaint yourself with the lovely summer constellations of **Hercules**, **Scorpius**, **Lyra**, **Cygnus** and **Aquila**.

▼ *The sky at 11 pm in mid-June, with Moon positions at three-day intervals either side of Full Moon. The star positions are also correct for midnight at the beginning of*

JUNE'S CONSTELLATION

Libra (the Scales) is a faint triangle of stars lying close to the constellation of **Scorpius**. In fact, the Greeks actually regarded its three brightest stars as the scorpion's claws.

The Arabs gave these stars delightful names. Alpha Librae (magnitude +2.8) is called **Zubenelgenubi**, meaning 'the southern claw'. It has a faint companion of magnitude +5.2, visible in good binoculars or a small telescope. The faintest of the three stars – magnitude +3.9 – is **Zubenelakrab** ('the scorpion's claw').

The brightest star in Libra is beta Librae – **Zubeneschamali** ('the northern claw'), which has a fascinating history. It now shines at a brightness of magnitude +2.6 (confusingly, brighter than alpha Librae!). But two Greek astronomer-philosophers of 2000 years ago rated it as brilliant as red **Antares** – the first-magnitude star that marks the scorpion's heart. Might this blue dwarf star have flared up in the past? There are suspicions of a companion star, which could have dumped material on beta Librae. In which case – watch that star in case it happens again!

Finally: something you won't be able to see for yourself. Libra boasts a faint star named Gliese 581, which has a family of at least three planets. From the gravitational 'wobbles' on the

Star chart labels

WEST

20 June
Regulus
Ecliptic
Jupiter
The Sickle
Venus
CANCER
LEO
NW
Pollux
Castor
GEMINI
URSA MAJOR
CANES VENATICI
AURIGA
The Plough
Capella
URSA MINOR
Polaris
DRACO
Zenith
HERCULES
NORTH
CASSIOPEIA
CEPHEUS
Vega
LYRA
PERSEUS
THE MILKY WAY
Algol
Deneb
CYGNUS
ANDROMEDA
DELPHINUS
NE
Square of Pegasus
PEGASUS
EAST

June, and 10 pm at the end of the month. The planets move slightly relative to the stars during the month.

star, one of them appears to be of relatively low mass – five times heavier than the Earth – and lives within its star's 'habitable zone', where life could be possible....

PLANETS ON VIEW

You'll find the glorious Evening Star low in the west, reaching its greatest elongation from the Sun on **6 June** – though **Venus** doesn't appear as spectacular as last month, as it's not in a totally dark sky. Starting the month near Castor and Pollux in Gemini, Venus races through Cancer and Leo towards slower-moving Jupiter. Through a small telescope, you'll see its shape change from half-lit to a fat crescent.

Meanwhile, **Jupiter** (at magnitude −0.7, brighter than any of the stars) lies to the left of Venus, and trundles slowly from Cancer into Leo. The chase ends when Venus catches up with the giant planet on **30 June**, passing just 20 arc minutes apart. They're closest together as the canoodling planets set in the west at 11 pm.

Saturn keeps a respectful distance in Libra, shining at magnitude +0.3 and setting around 3.30 am.

Faint **Neptune**, at magnitude +7.9, lies in Aquarius and rises at 1 am. It's followed by fellow giant **Uranus** (magnitude +5.9), rising around 2 am in Pisces.

Mercury is too low in the morning twilight to be seen from Britain this month, although it reaches greatest western elongation on **24 June**. **Mars**, too, is hidden in the Sun's glare in June.

WEST

EAST

Ecliptic

23 June

LEO

VIRGO

Spica

CORVUS

HYDRA

SW

URSA MAJOR

CANES VENATICI

BOÖTES

Arcturus

26 June

Zubenelgenubi

LIBRA

The Plough

DRACO

Zenith

M13

CORONA BOREALIS

SERPENS

M5

Zubeneschamali

Zubenelakrab

Saturn

Antares

SCORPIUS

SOUTH

Vega

LYRA

HERCULES

OPHIUCHUS

2 June

Ecliptic

SAGITTA

CYGNUS

SERPENS

SAGITTARIUS

THE MILKY WAY

PEGASUS

DELPHINUS

Altair

AQUILA

CAPRICORNUS

AQUARIUS

SE

June's Object
Vega

June's Picture
M5

Venus

Jupiter

Saturn

Moon

MOON		
Date	Time	Phase
2	5.19 pm	Full Moon
9	4.42 pm	Last Quarter
16	3.05 pm	New Moon
24	12.03 pm	First Quarter

MOON

The almost Full Moon lies right next to Saturn on **1 June**. The crescent forms one corner of a celestial triangle on **20 June**, along with brilliant Venus and Jupiter. On **25 June**, the waxing Moon is to the right of Spica. The Moon returns to Saturn on **28 June**.

SPECIAL EVENTS

21 June, 5.38 pm: Summer Solstice. The Sun reaches its most northerly point in the sky, so 21 June is Midsummer's Day, with the longest period of daylight. Correspondingly, we have the shortest nights.

JUNE'S OBJECT

One of our favourite stars – **Vega**, in **Lyra** – is rapidly ascending the heavens to occupy the zenith for the summer. The fifth brightest star in the sky, Vega has the honour to have been the first star photographed after the Sun. It is pure white: so pure that its colour is used as a benchmark to measure the colours of other stars – from red to blue-white – and so gauge their temperatures.

The star is a whirling dervish. It rotates in just 12.5 hours (as compared to roughly 30 days for our Sun), and – as a result – its equator bulges outwards, making Vega a satsuma-shaped star.

▲ *This image of globular cluster M5 was photographed by Luke Broom-Lynne from Attleborough, Norfolk. It was captured on 3 May 2011 with a Canon 450D DSLR attached to a 250 mm f/4.7 reflecting telescope equipped with a light-pollution filter. The 39 separate 5-minute exposures were stacked in Registax 4 software, and processed in Photoshop.*

◉ Viewing tip

June is *the* month for the best Sun-viewing, and this year is a great time to catch up with our local star, as its sunspot activity (and associated flares and eruptions) is near its (roughly) 11-year maximum. But be careful. **NEVER** use a telescope or binoculars to look at the Sun directly: it could blind you permanently. Fogged film is no safer, because it allows the Sun's infrared (heat) rays to get through. Eclipse goggles are safe (unless they're scratched). The best way to observe the Sun is to project its image through binoculars or a telescope on to a white piece of card. Or – if you want the real 'biz' – get some solar binoculars or even a solar telescope, with filters that guarantee a safe view. Check the web or a specialist equipment supplier for details.

Vega was one of the first stars around which astronomers discovered a warm, dusty disc – like the one which formed our planetary system 4.6 billion years ago. There are strong suspicions that a planet the mass of Jupiter might be lurking in there.

JUNE'S PICTURE

Messier 5 (**M5**) in **Serpens** is a giant amongst globular clusters (see Topic below). Just visible to the unaided eye in really transparent skies (it's next to the faint star 5 Serpentis – left in the image), it shines at magnitude +6.6. One of the biggest globular clusters known, M5 measures 165 light years across, and contains between 100,000 and 500,000 stars – no one's certain! Ever a record-breaker, M5 is very distant: 24,500 light years away. And at 13 billion years old, it's very ancient. To cap it all, the cluster is teeming with variable stars – over 100 have now been logged.

JUNE'S TOPIC
Globular clusters

Look the right of **Vega**, and slightly downwards, and you'll spot a faint rectangle of stars – the body of the superhero **Hercules**. Drop down from the top right-hand star, and you'll see a fuzzy patch (binoculars will help). This is **M13**: a ball of around a million stars.

M13 is just one of over 150 globular clusters that swarm around our Milky Way and a cousin to M5 (see this month's Picture). All major galaxies are surrounded by globular clusters – the more massive the galaxy, the bigger the swarm. The giant galaxy M87 in Virgo may be circled by 13,000 of these ancient artefacts.

Globular clusters are very old: their stars are elderly red giants, and – unlike the sparkling open clusters, which grace our Galaxy's disc today – these giant clusters show no signs of star formation.

The stars in a globular cluster are densely packed. The average separation between their residents is just one light year (for comparison, we are 4.3 light years from the Sun's nearest neighbour, Proxima Centauri). This close packing makes globular clusters an unlikely home for alien life. Any planetary systems that formed here would have been disrupted by the gravitational pull of stars passing nearby.

Nonetheless, globular clusters are a glorious sight in binoculars or a small telescope – check them out on our list of deep-sky objects (pages 58–60).

Venus and Jupiter have dominated our nights all year – but, amazingly, by the end of July they've both disappeared from the evening sky. Instead, seek out some of the stellar delights of high summer. The brilliant trio of the **Summer Triangle** – the stars **Vega**, **Deneb** and **Altair** – is composed of the brightest stars in the constellations **Lyra**, **Cygnus** and **Aquila**. And this is the time to catch the far-southern constellations of **Sagittarius** and **Scorpius** – embedded in the glorious heart of the Milky Way.

▼ The sky at 11 pm in mid-July, with Moon positions at three-day intervals either side of Full Moon. The star positions are also correct for midnight at the beginning of

JULY'S CONSTELLATION

Down in the deep south of the sky this month lies a baleful red star. This is **Antares** – 'the rival of Mars' – and in its ruddiness it even surpasses the famed Red Planet. To ancient astronomers, Antares marked the heart of **Scorpius**, the celestial scorpion.

According to Greek myth, this summer constellation is intimately linked with the winter star pattern Orion, who was killed by a mighty scorpion. The gods immortalized these two opponents as star patterns, set at opposite ends of the sky so that Orion sets as Scorpius rises.

Scorpius is one of the few constellations that look like their namesakes. To the top right of Antares, a line of stars marks the scorpion's forelimbs. Originally, the stars we now call **Libra** (the Scales) were its claws. Below Antares, the scorpion's body stretches down into a fine curved tail (below the horizon on the chart), and deadly sting. Alas – the curved sting isn't visible from latitudes as far north as the UK. A good excuse for a Mediterranean holiday!

Scorpius is a treasure-trove of astronomical goodies. Several lovely double stars include Antares: its faint companion looks greenish in contrast to Antares' strong red hue. Binoculars reveal the fuzzy patch of **M4**, a globular cluster

July, and 10 pm at the end of the month. The planets move slightly relative to the stars during the month.

made of tens of thousands of stars, some 7200 light years away.

The 'sting' contains two fine star clusters – **M6** and **M7** – so near to us that we can see them with naked eye: a telescope reveals their stars clearly.

PLANETS ON VIEW

The almost conjoined twins of **Venus** (magnitude −4.3) and **Jupiter** (magnitude −1.6) – only half a degree apart on **1 July** – start the month very low in the north-west after sunset, and quickly slip downwards to disappear by the end of the month. Through a telescope, Venus appears as a lovely crescent.

Saturn has the later evening sky all to itself. At magnitude +0.4, the ringed planet inhabits the dim constellation of Libra, and sets around 1.30 am.

At a measly magnitude +7.8, **Neptune** rises at 11 pm in Aquarius, while **Uranus** (magnitude +5.8) follows in Pisces, popping above the horizon at midnight.

Mercury and **Mars** are too close to the Sun to be visible this month.

MOON

The thinnest crescent Moon appears below Venus and Jupiter soon after sunset on **18 July** – best seen with binoculars. On **23 July**, the Moon lies near Spica. You'll find Saturn to the left of the Moon on **25 July**, and to its right on **26 July**.

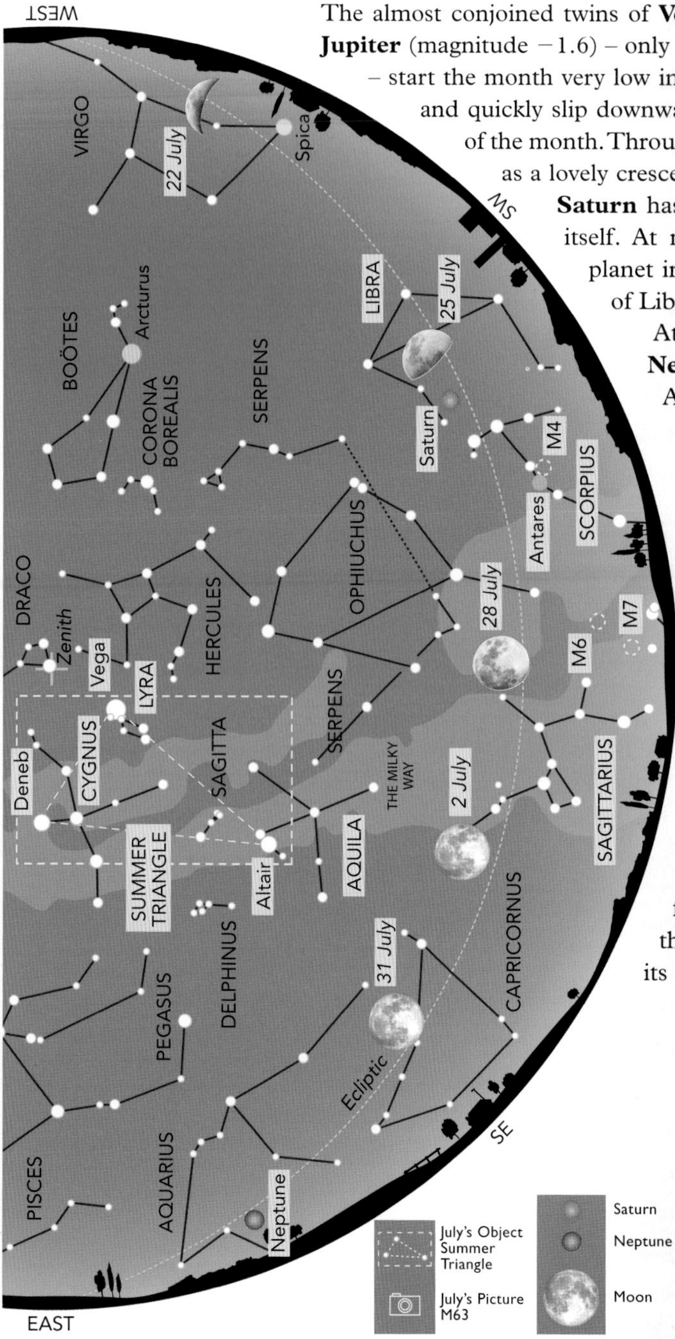

MOON		
Date	**Time**	**Phase**
2	3.19 am	Full Moon
8	9.24 pm	Last Quarter
16	2.24 am	New Moon
24	5.04 am	First Quarter
31	11.43 am	Full Moon

WEST

VIRGO · 22 July · Spica · WSW · BOÖTES · Arcturus · CORONA BOREALIS · SERPENS · LIBRA · 25 July · Saturn · M4 · SCORPIUS · Antares · M7 · M6 · 28 July · OPHIUCHUS · SERPENS · HERCULES · DRACO · Zenith · Vega · LYRA · CYGNUS · Deneb · SAGITTA · SUMMER TRIANGLE · AQUILA · Altair · DELPHINUS · THE MILKY WAY · 2 July · SERPENS · SAGITTARIUS · SOUTH · 31 July · CAPRICORNUS · Ecliptic · SE · PEGASUS · DELPHINUS · AQUARIUS · Neptune · PISCES · EAST

July's Object
Summer Triangle

July's Picture
M63

Saturn
Neptune
Moon

31

SPECIAL EVENTS

On **6 July**, at 8.40 pm, the Earth reaches aphelion, its furthest point from the Sun – 152 million kilometres out.

The New Horizons spacecraft will whiz past Pluto on **14 July** (see this month's Topic).

JULY'S OBJECT

The **Summer Triangle** is very much part of this season's skies (and it hangs around for most of the autumn, too!). It's made up of Vega, Deneb and Altair – the brightest stars in the constellations of **Lyra** (the Lyre), **Cygnus** (the Swan) and **Aquila** (the Eagle), respectively. The trio of stars make a striking pattern almost overhead on July nights.

The stars may seem to be almost the same brightness, but they're very different beasts. **Altair** – its name means 'flying eagle' – is one of the Sun's nearest neighbours, at a distance of 17 light years. It's 11 times brighter than the Sun and spins at a breakneck rate of once every nine hours – as compared to around 30 days for our local star.

Vega, just over 25 light years away, is a brilliant white star nearly twice as hot as the Sun. In 1850, it was the first star to be photographed. Now, more sensitive instruments have revealed that Vega is surrounded by a dusty disc – which may be a planetary system in the process of formation.

While **Deneb** – meaning 'tail' (of the swan) – may appear to be the faintest of the trio, the reality is different. It lies at least 1500 light years away (the Hipparcos satellite puts it as far off as 3200 light years) – because the star is so distant, these measurements will always be controversial, so watch this space! To appear so bright in our skies, it must be truly luminous. It's estimated that Deneb is over 60,000 times brighter than our Sun – one of the most brilliant stars known.

▼ *Steve Allan captured this image of M63 from a backyard in Sandbach, Cheshire. His telescope was a 190 mm Sky-Watcher f/5.3 Maksutov-Newtonian, coupled to a QHY9 Mono camera with 2-inch LRGB filters. The total exposure time – over three nights in April 2011 – was 14.5 hours. The final image was processed in MaximDL, DeepSkyStacker and Photoshop.*

JULY'S PICTURE

The obscure constellation of **Canes Venatici** (the Hunting Dogs) lies just below the 'tail' of the Great Bear. It's home to about a dozen galaxies, which form a little group. The most spectacular is the glorious Whirlpool Galaxy, M51 – but this stunning cosmic vortex has its rivals. One member of the cluster, located about 40 million light years away, is the Sunflower Galaxy, **M63**. In the mid-19th century, Lord Rosse – in Ireland – discovered its spiral arms, using the biggest telescope in the world (with a mirror 6 feet across). There was little follow-up, though – astronomers at the time didn't know what galaxies actually were. In 1971, a supernova appeared in one of the galaxy's spiral arms.

◉ Viewing tip

This is the month when you really need a good, unobstructed view of the southern horizon to see the summer constellations of Scorpius and Sagittarius. They never rise high in temperate latitudes, so make the best of a southerly view – especially over the sea – if you're away on holiday. A good southern horizon is also best for views of the planets, because they rise highest when they're in the south.

JULY'S TOPIC
New Horizons at Pluto

Pluto has long been every kids' favourite planet. Discovered in 1930 by American astronomer Clyde Tombaugh, this tiny world has gripped the imagination of all and sundry.

Until it became evident that Pluto was not alone. It is a member of a swarm of 'Kuiper Belt Objects': minuscule worlds at the edge of our Solar System. And in 2005, a team led by Mike Brown – observing from Palomar Mountain in California – discovered a distant object even more massive (and possibly slightly larger) than Pluto.

The new body was named Eris – a Greek goddess of discord and strife. Her name would prove to be prophetic, because her discovery challenged the status of Pluto. In August 2006, at a meeting of the International Astronomical Union in Prague, Pluto's ranking in the Solar System was downgraded to that of 'dwarf planet' – much to the disappointment of Pluto *aficionados*. Now there are officially only eight planets in the Solar System.

But fans of the mini-world have an event to look forward to in 2015. In July, the space probe New Horizons will arrive at the Pluto system after a nine-year journey through space. It will pass within 10,000 kilometres of the dwarf planet, and 27,000 kilometres from Pluto's major moon, Charon.

After studying the geology and chemistry of Pluto and its moons – the mini-world has at least five in total – New Horizons will be ready to move on. It is destined to explore an uncharted zone of our Solar System: the frozen wasteland of the Kuiper Belt Objects – a twilight zone where no spacecraft has ever been before.

After the glut of planets so far this year, August is pretty much a desert for anyone interested in the other worlds of the Solar System. To make up for this, we are in for a fabulous display of celestial fireworks – the famous **Perseid** meteors, which peak on the night of **12/13 August**.

▼ The sky at 11 pm in mid-August, with Moon positions at three-day intervals either side of Full Moon. The star positions are also correct for midnight

AUGUST'S CONSTELLATION

It has to be admitted that **Aquila** does vaguely resemble a flying eagle, albeit a rather faint one. It's an ancient constellation, named after the bird which was a companion to the god Jupiter – and even carried his thunderbolts for him!

The constellation is dominated by **Altair**, a young blue-white star 17 light years away, which is 11 times brighter than our Sun. It has a very fast spin: the star hurtles around at 210 kilometres per second, rotating in just nine hours (as opposed to about 30 days for the Sun). As a result, it is oval in shape.

Altair is a triple star, as is its neighbour **beta Aquilae** (just below, left). **Eta Aquilae** (immediately below beta) is one of the brightest Cepheid variable stars – old stars that change their brightness by swelling and shrinking. The pulsations of eta Aquilae make it vary from magnitude +3.5 to +4.4 every seven days.

PLANETS ON VIEW

Saturn is the sole bright planet in the early evening sky, setting about 11.30 pm. Lying in Libra, Saturn (magnitude +0.6) is the brighter of the two objects on view low in the south-west, the other being red giant Antares (magnitude +1.0). Grab a telescope if you can, to view Saturn's splendid rings.

Neptune is visible all night long in Aquarius – though, at magnitude +7.8, you'll need a telescope to spot it. **Uranus** (magnitude +5.8) is rising around 10 pm, in Pisces.

at the beginning of August, and 10 pm at the end of the month. The planets move slightly relative to the stars during the month.

In the dawn sky, **Mars** is rising at 4 am, and becomes more prominent during the month as sunrise becomes later. The Red Planet glows at magnitude +1.7, in Cancer. Mars lies in front of the Beehive star cluster on the mornings of **20 and 21 August** (use binoculars or a small telescope for the best view).

Venus, at magnitude −4.2, reappears as the Morning Star during the last few days of August, low in the east before sunrise.

Mercury and **Jupiter** are lost in the Sun's glare throughout August.

MOON

The crescent Moon lies above Spica on **19 August**. And you'll find Saturn below the First Quarter Moon on **22 August**.

SPECIAL EVENTS

The maximum of the **Perseids** falls on **12/13 August**. It's an excellent year for observing one of the most prolific annual meteor showers, as the Moon is well out of the way. Expect a high proportion of fast bright shooting stars.

Comet Churyumov–Gerasimenko is closest to the Sun on **13 August**: it's too faint and low in the morning glow for easy observation from Earth, but – if all has gone to plan – there should be amazing close-up views from the orbiting Rosetta spacecraft and the Philae lander, anchored to the comet's surface.

WEST

EAST

LIBRA · **SERPENS** · **CORONA BOREALIS** · **SCORPIUS** · **Saturn** · **OPHIUCHUS** · 23 Aug · MS · **HERCULES** · **DRACO** · Vega · **LYRA** · **SAGITTA** · Altair · **AQUILA** · eta · beta · THE MILKY WAY · **SERPENS** · **SAGITTARIUS** · SOUTH · 26 Aug · Zenith · Deneb · **CYGNUS** · THE GREAT RIFT · **DELPHINUS** · **CAPRICORNUS** · **ANDROMEDA** · **PEGASUS** · 29 Aug · Neptune · **PISCIS AUSTRINUS** · Square of Pegasus · **PISCES** · 2 Aug · **AQUARIUS** · SE · 5 Aug · Uranus · Ecliptic · **CETUS**

		MOON	
	Date	**Time**	**Phase**
Saturn	7	3.03 am	Last Quarter
Uranus	14	3.53 pm	New Moon
Neptune	22	8.31 pm	First Quarter
Moon	29	7.35 pm	Full Moon

August's Object
The Milky Way

Radiant of
Perseids

35

On **15 August**, the European Space Agency will launch the BepiColombo spacecraft to orbit Mercury; it should arrive in January 2022.

AUGUST'S OBJECT

It's a stunning month for sweeping down **the Milky Way**, especially through binoculars. The stars look packed together, and you'll pick out star clusters and nebulae as you travel its length. These are all the more distant denizens of our local Galaxy, flattened into a plane because we live within its disc. It's akin to seeing the overlapping streetlights of a distant city on Earth.

But you'll notice something else – there is a black gash between the stars. William Herschel, the first astronomer to map the Galaxy, thought that this was a hole in space. But now we know that **the Great Rift** in **Cygnus** is a dark swathe of sooty dust crossing the disc of our Galaxy. It is material poised to collapse under gravity, heat up, and – mixed with interstellar gas – it will create new generations of stars and planets. Proof that there is life in our old Galaxy yet!

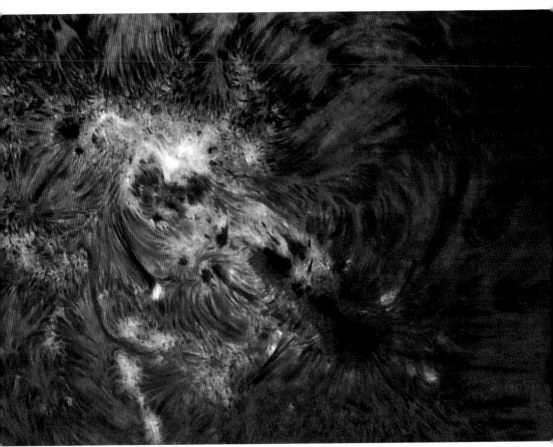

AUGUST'S PICTURE

The **Sun** is still going through its peak of solar activity, when its magnetic field gets periodically wrapped up around it like an elastic band. The result is an outbreak of sunspots: cooler regions on the Sun's surface where its churning gases are dammed back. Sunspots appear dark against the Sun's brilliant disc, and they can be vast – much bigger than our planet. When the suppressed energy is finally released, the Sun bursts into serious eruptions – solar flares and coronal mass ejections – whose electrical particles can affect life on Earth.

This image is taken through a filter that isolates the light from glowing hydrogen in the lower chromosphere, the layer of the Sun's atmosphere just above the visible surface. Around the dark sunspot (lower right), loops of magnetic field are corralling the gas into narrow bands. Concentrated magnetic energy is creating a hot-spot (upper left) in the chromosphere, called a plage (from the French for 'beach').

AUGUST'S TOPIC
Pulsars

Between **Sagitta** and **Cygnus**, there's a constellation so obscure it's not even marked on our chart: Vulpecula (the Fox). But it has a unique place in astronomical history.

Here, in 1967, Cambridge astronomers found an object broadcasting a regular stream of radio pulses, once every 1.337 seconds. At first, they thought it might be a signal from alien intelligence, and dubbed it LGM-1 (for Little Green Men). In fact, it was the first pulsar.

We now know a pulsar is the collapsed core of an old star that has exploded as a supernova. It's composed entirely of tiny subatomic particles called neutrons, so tightly packed that a pulsar (also called a neutron star) contains as much matter as the Sun, in a ball no bigger than London!

A pinhead of its material would weigh as much as a fully laden supertanker. And the pulsar's gravity is so strong you'd expend more effort climbing a 1-centimetre hump on its surface than in ascending Mount Everest on Earth.

A typical pulsar has a fearsome magnetic field, a thousand billion times stronger than Earth's magnetism. As it spins round, beams of radiation from the pulsar's magnetic poles sweep around like a lighthouse beacon, creating the radio pulses we detect from Earth.

In the case of LGM-1, the observations show it's spinning round in just over a second. The most extreme pulsar, known as J1748-2446ad, is rotating an amazing 716 times per second – faster than a kitchen blender!

◀ *Stuart Green captured this image of a sunspot from his back garden in Preston using an Etalon solar filter tuned to the red light of hydrogen-alpha. His instrument was a 150 mm f/10 telescope. This vivid image captures the power and fury of a sunspot.*

This month's highlight is the best total eclipse of the Moon we've seen from the UK since 2008 – it's well worth staying up until the wee small hours for the spectacle of a larger-than-average 'supermoon' sliding into Earth's shadow and turning blood-red.

Also, this month, the nights become longer than the days, as the Sun migrates southwards in the sky. Autumn is here – with its unsettled weather – and we have wet star patterns to match! **Aquarius** (the Water Carrier) is part of a group of aqueous star patterns which include **Cetus** (the Sea Monster), **Capricornus** (the Sea Goat), **Pisces** (the Fishes), **Piscis Austrinus** (the Southern Fish) and **Delphinus** (the Dolphin).

SEPTEMBER'S CONSTELLATION

Piscis Austrinus (the Southern Fish) lies low in the south-west. It's hardly a compelling constellation. But it has fond childhood memories for both of us, because this is the only time of year that we got to see its brightest star, **Fomalhaut**: the south-ernmost first-magnitude star visible from Britain.

Its name, which is derived from the Arabic, means 'the mouth of the whale', though one of its nicknames seems more appropriate: 'the lonely star of autumn'.

But Fomalhaut isn't really lonely. It has a disc of debris in orbit about it, with the potential to form a planetary family. And in 2008, the Hubble Space Telescope captured an image of one of its worlds – the first to be seen beyond the Solar System.

PLANETS ON VIEW

In the evening sky, look out for **Saturn**, low down in the south-west in Libra. At magnitude +0.7, the ringworld is marginally brighter than the star Antares, to its left, and sets around 9.45 pm.

▼ The sky at 11 pm in mid-September, with Moon positions at three-day intervals either side of Full Moon. The star positions are also correct for midnight at

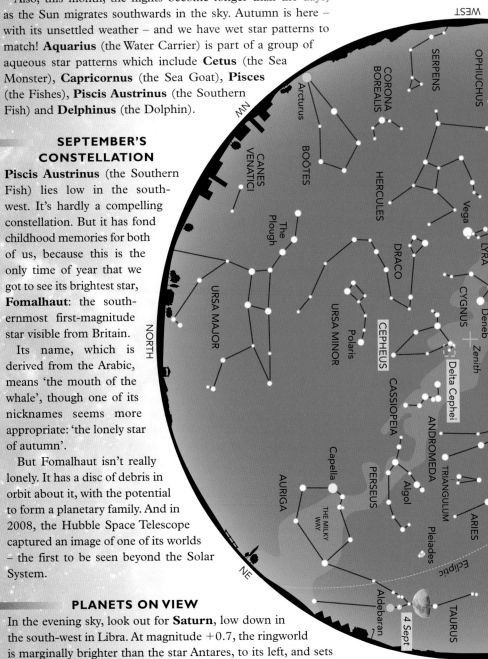

the beginning of September, and 10 pm at the end of the month. The planets move slightly relative to the stars during the month.

The most distant planet, **Neptune**, is at opposition on **1 September** and at its closest to Earth – though you'll still need a telescope to spot the magnitude +7.8 world. Neptune lies above the horizon all night long, in Aquarius.

Uranus, at magnitude +5.7 in Pisces, rises around 9 pm.

We have a brilliant Morning Star! **Venus** is speeding upwards in the dawn sky, blazing at magnitude −4.4 in the east. By the end of September, it's rising almost four hours before the Sun. Through a telescope, watch Venus's narrow crescent fill out to almost half-lit as the month progresses.

Mars lies well to the left of Venus, but is a dim shadow of its sister: 300 times fainter, at magnitude +1.8. Rising just before 4 am, the Red Planet travels from Cancer into Leo, and passes very close to its brightest star, Regulus, on the morning of **25 September**: there's a lovely colour contrast (best seen in binoculars) between the famously ruddy Mars and the blue-white star.

Following behind Mars, you will find brilliant **Jupiter** (magnitude −1.6) moving upwards in the dawn twilight, and rising by 4.30 am at the end of the month.

Mercury lies too close to the Sun to be visible from Britain this month, even though it's at greatest eastern elongation on **4 September**.

MOON

Just before dawn on **5 September**, the Moon moves right in front of Aldebaran (see Special Events). The thin crescent lies next to Venus on the morning of

Star chart constellation and star labels: WEST, EAST, SE, SOUTH, SERPENS, OPHIUCHUS, HERCULES, SERPENS, THE MILKY WAY, Vega, LYRA, SAGITTA, AQUILA, Altair, DELPHINUS, SAGITTARIUS, 22 Sept, CAPRICORNUS, CYGNUS, Deneb, Zenith, CEPHEUS, Delta Cephei, ANDROMEDA, PEGASUS, Square of Pegasus, AQUARIUS, 25 Sept, Neptune, PISCIS AUSTRINUS, GRUS, Fomalhaut, Ecliptic, TRIANGULUM, PISCES, 28 Sept, Uranus, Mira, CETUS, ARIES, TAURUS, ERIDANUS

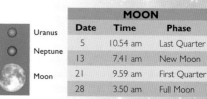

			September's Object Delta Cephei		

Uranus
Neptune
Moon
September's Object Delta Cephei
September's Picture Total lunar eclipse

MOON		
Date	**Time**	**Phase**
5	10.54 am	Last Quarter
13	7.41 am	New Moon
21	9.59 am	First Quarter
28	3.50 am	Full Moon

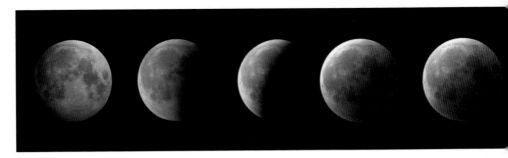

10 September, with Mars to the left. Saturn is near the Moon on **18 September**.

SPECIAL EVENTS

On the morning of **5 September**, the Last Quarter Moon passes through the Hyades star cluster, culminating in an occultation of Aldebaran at some time between 5.40 am and 5.55 am, depending on your location in Britain.

On **13 September**, a partial eclipse of the Sun will be visible from southern Africa, Madagascar, parts of Antarctica and the southern Indian Ocean, but not from Britain.

It's the Autumn Equinox at 9.20 am on **23 September**. The Sun is over the Equator as it heads southwards in the sky, and day and night are equal.

The night of **27/28 September** sees a total eclipse of the Moon that's visible from Europe (including Britain), western Asia, Africa, the Pacific Ocean and the Americas. It's also the night of a 'supermoon' (see this month's Topic).

SEPTEMBER'S OBJECT

At first glance, the star **Delta Cephei** – in the constellation representing King **Cepheus** – doesn't seem to merit any special attention. It's a yellowish star of magnitude +4: easily visible to the naked eye, but not prominent. A telescope reveals a companion star. But this star holds the key to measuring the size of the Universe.

Check the brightness of this star carefully over days and weeks, and you'll see that its brightness changes regularly, from +3.5 (brightest) to +4.4 (faintest), every 5 days 9 hours. It's a result of the star literally swelling and shrinking in size, from 32 to 35 times the Sun's diameter.

Astronomers have found that stars like this – Cepheid variables – show a link between their period of variation and their intrinsic luminosity. By observing the star's period and brightness as it appears in the sky, astronomers can work out a Cepheid's distance. With the Hubble Space Telescope, astronomers have now measured Cepheids in the galaxy NGC 4603, which lies 100 million light years away.

⊙ **Viewing tip**

Now that the nights are drawing in earlier, and becoming darker, it's a good time to pick out faint, fuzzy objects such as star clusters, nebulae and galaxies. But don't even think about it near the time of Full Moon – its light will drown them out. The best time to observe 'deep-sky objects' is just before or after New Moon. Check the Moon phases timetables in the book.

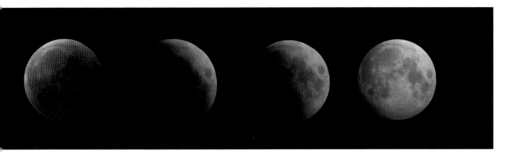

▲ *James Jefferson captured this series of images of a lunar eclipse from Ruislip, Middlesex. You can see from the shadow of the Earth on the Moon that our world is four times wider than our companion. He used a Celestron 9.5 telescope, coupled to a Nikon 4300 Digital Camera, and enhanced the images of totality to bring out the reddish Earth-glow on our satellite.*

SEPTEMBER'S PICTURE

This year, we're treated to the best **total eclipse of the Moon** for several years (see this month's Topic). It's not crucial to be in a particular place: lunar eclipses favour half the world, and – this time – it's our turn. Just set your alarm for the early hours of the morning!

SEPTEMBER'S TOPIC
Eclipse of the supermoon

We're treated to a particularly big and bright Full Moon this month. On the night of **27/28 September**, our companion world is at its closest point this year – just 356,900 kilometres away – and appears 14% bigger than when the Moon is farthest away. By coincidence the Moon is Full just an hour later, so the brilliant orb will be extra-bright – some 30% more luminous than the faintest Full Moon.

A Full Moon at the closest point in its orbit (perigee) is called a 'supermoon'. We're afraid to say that this catchy phrase wasn't invented by astronomers, but by astrologers who have tried to link supermoons with earthquakes, tsunamis or volcanic eruptions. But the supermoon's extra gravitational pull is only 3% more powerful than its tug on the Earth when the Moon reaches its average perigee distance in each monthly orbit – so we can confidently predict the supermoon will cause no natural calamities.

And this supermoon is a bit special – it coincides with a total lunar eclipse. The Moon starts to slide into the Earth's shadow at 2.07 am; it's at perigee at 2.46 am; and is totally eclipsed at 3.11 am. Though the Moon is in the Earth's shadow, some sunlight is bent by the Earth's atmosphere to illuminate the Moon in a faint coppery glow. We can't predict the darkness of the eclipse, as it depends on the cloudiness of the Earth's atmosphere.

But the supermoon phenomenon provides the unusual sight of the most brilliant Full Moon of the year quickly descending to near invisibility.

The glories of October's skies can best be described as 'subtle'. The barren square of **Pegasus** dominates the southern sky, with **Andromeda** attached to his side. But the dull autumn constellations are already being challenged by the brilliant lights of winter, spearheaded by the beautiful star cluster of the **Pleiades**. From Greece to Australia, ancient myths independently describe these stars as a group of young girls being chased by an aggressive male – often **Aldebaran** or **Orion**.

OCTOBER'S CONSTELLATION

Pegasus is little more than a large, empty square of four medium-bright stars. But our ancestors somehow managed to see the shape of an upside-down winged horse here. In legend, Pegasus sprang from the blood of Medusa the Gorgon when **Perseus** (nearby in the sky) severed her head.

The star at the top right of the square – **Scheat** – is a red giant over a hundred times wider than the Sun. Close to the end of its life, it pulsates irregularly, changing in brightness by about a magnitude. Outside the square, **Enif** ('the nose') is a yellow supergiant. A small telescope, or even good binoculars, reveals a faint blue companion star.

Just next to Enif – and Pegasus's best-kept secret – is the beautiful globular cluster **M15**. You'll need a telescope for this one. M15 is around 30,000 light years away, and contains about 100,000 densely packed stars.

PLANETS ON VIEW

Sultry **Saturn** (magnitude +0.7) – still haunting Libra – lies low in the south-west after sunset. It sinks into the twilight glow at the end of this month.

If you like a challenge, check out the other two evening planets. You'll need good binoculars or a telescope to track

▼ The sky at 11 pm in mid-October, with Moon positions at three-day intervals either side of Full Moon. The star positions are also correct for midnight at

WEST

OPHIUCHUS
AQUILA
CORONA BOREALIS
HERCULES
LYRA
THE MILKY WAY
Vega
CYGNUS
BOÖTES
DRACO
Deneb
CANES VENATICI
The Plough
CEPHEUS
Zenith
CASSIOPEIA
URSA MINOR
Polaris
NORTH
URSA MAJOR
PERSEUS
Capella
AURIGA
Castor
GEMINI
Pollux
Radiant of Orionids
Ecliptic
30 Oct
Aldebaran
Betelgeuse
ORION
NE

the beginning of October, and 9 pm at the end of the month (after the end of BST). The planets move slightly relative to the stars during the month.

down **Neptune**, at magnitude +7.8 in Aquarius. It sets around 3 am.

Uranus is at opposition on **12 October** and closest to the Earth. At magnitude +5.7, it's just visible to the naked eye, but binoculars will help you find this distant world, above the horizon all night long in Pisces.

In the early hours of the morning, we have a planetary waltz in Leo. First on the stage is brilliant **Venus** (magnitude −4.3), rising at 3 am at the beginning of October. The Evening Star is followed by **Mars**, 250 times fainter at magnitude +1.7, and moving above the horizon at 3.45 am. (The star Regulus lies between Mars and Venus.) Then comes **Jupiter**, the second brightest planet (magnitude −1.6), which rises around 4.30 am.

As the month progresses, both Venus and Mars speed down towards Jupiter. Venus passes Regulus on **9 October**, with the crescent Moon nearby (see Special Events). On **18 October**, Mars skims less than half a degree from Jupiter. And, on **26 October**, Venus – at greatest western elongation – flies under Jupiter.

To complete the planetary roster, Mercury makes its best morning appearance of the year, reaching greatest western elongation on **16 October**. You can spot the innermost planet very low in the east, around 6 am, from **11 October** to almost the end of the month, as its magnitude soars from +0.4 to −0.9.

WEST

SERPENS
THE MILKY WAY
AQUILA
Altair
SAGITTA
CYGNUS
Deneb
DELPHINUS
M15
Enif
PEGASUS
Scheat
Square of Pegasus
CASSIOPEIA
Zenith
ANDROMEDA
PERSEUS
TRIANGULUM
Pleiades
ARIES
PISCES
Uranus
Mira
TAURUS
Aldebaran
Betelgeuse
ORION
Rigel

AQUARIUS
Ecliptic
Neptune
CAPRICORNUS
21 Oct
MS
PISCIS AUSTRINUS
Fomalhaut
24 Oct
CETUS
27 Oct
ERIDANUS
SE
SOUTH

EAST

MOON		
Date	**Time**	**Phase**
4	10.06 pm	Last Quarter
13	1.06 am	New Moon
20	9.31 pm	First Quarter
27	12.05 pm	Full Moon

Uranus

Neptune

October's Object
Uranus

Radiant of
Orionids

Moon

MOON

On the morning of **8 October**, the crescent Moon is to the upper right of Venus; on **9 October** it's near Venus, Regulus, Mars and Jupiter (see Special Events); and on **10 October** you'll find the Moon below Jupiter. Before dawn on **11 October**, the slim crescent Moon lies just to the right of Mercury, very low in the east (use binoculars for the best view). Back in the evening sky, the crescent Moon passes above Saturn on **16 October**. On **29 October**, the Moon occults Aldebaran (see Special Events).

SPECIAL EVENTS

On **9 October**, we have a splendid tableau in the morning sky, when the crescent Moon joins the two brightest planets – Venus and Jupiter – along with Red Planet Mars and the blue-white star Regulus.

Debris from Halley's Comet smashes into Earth's atmosphere on **21/22 October**, causing the annual **Orionid** meteor shower. They'll be best seen in the early hours of the morning, after the Moon has set.

At 2 am on **25 October**, we see the end of British Summer Time for this year. Clocks go backwards by an hour.

On **29 October**, the Moon moves in front of Aldebaran, at a convenient time in the evening – somewhere between 9.40 pm and 10.00 pm depending on your location in Britain. The occultation lasts about an hour.

OCTOBER'S OBJECT

If you're very sharp-sighted and have extremely dark skies, you stand a chance of spotting **Uranus** – the most distant planet visible to the unaided eye. At magnitude +5.7 in Pisces, it's closest to Earth this year on **12 October**. Discovered in 1781 by amateur astronomer William Herschel, Uranus was the first planet to be found since antiquity. Then, it doubled the size of our Solar System.

Four times the diameter of the Earth, Uranus has an odd claim to fame: it orbits the Sun on its side (probably as a result of a collision in its infancy). Like the other gas giants, it has an encircling system of rings. But these are nothing like the spectacular edifices

▼ *Before its fatal demise, Lawrence Dunn photographed Comet ISON – complete with its magnificent tail – from a car park in the New Forest, Hampshire. The image, captured in the dawn sky, was taken with an Atik 460EX camera, attached to a Takahashi FSQ108 f/5 telescope.*

⊙ Viewing tip

When you first go out to observe, you may be disappointed at how few stars you can see in the sky. But wait for around 20 minutes and you'll be amazed at how your night vision improves. One reason for this 'dark adaption' is that the pupil of your eye gets larger to make the best of the darkness. More importantly, in dark conditions the retina of your eye builds up much bigger reserves of rhodopsin, the chemical that responds to light.

that girdle Saturn – Uranus's 13 rings are thin and faint. It also has a large family of moons: at the last count, 27.

The giant planet consists largely of a vast watery ocean surrounding a hot rocky core. The latest theories posit a hailstorm of diamonds dropping down through the water, to land in a sea of liquid diamond that wraps around the core. Though the Voyager space probe revealed a bland, featureless world when it flew past Uranus in 1986, things are now hotting up as the planet's seasons change, with streaks and clouds appearing in its atmosphere.

OCTOBER'S PICTURE

Catching all the headlines late in 2013, **Comet ISON** was predicted to become the 'Comet of the Century'. Some pundits even estimated it would shine as brightly as the Full Moon. And images of the approaching comet were starting to look spectacular.

But in November 2013, ISON had its close brush with our local star. The result? Sun: 1 – ISON: 0. The Sun's heat and mighty gravity tore the fragile comet apart.

OCTOBER'S TOPIC
The Oort-Öpik Cloud

Everyone is thrilled when a new comet hoves into view. Even if the recent Comet ISON bombed, who can forget the magnificence of Hale-Bopp in 1997?

But, throughout history, comets have had a dodgy reputation. As Shakespeare wrote in *Julius Caesar*: 'When beggars die, there are no comets seen. The heavens themselves blaze forth the death of princes.' Looking like malevolent daggers hanging in the sky, comets were portents of disaster. And sometimes, these instincts were right. About 65 million years ago, a comet (or a small asteroid) collided with the Earth and destroyed the dinosaurs.

In reality, comets are 'dirty snowballs' – icy and rocky debris left over from the creation of the planets. Sweeping in towards the Sun, their ices boil, creating a massive head of gas and a trailing tail millions of kilometres long. But these maligned visitors may also be bearers of life, possibly delivering the ice that melted to become Earth's oceans – along with organic molecules.

Their homeland is a huge sphere surrounding our Solar System, named the Oort-Öpik Cloud after the two astronomers who deduced its existence. When a passing star disturbs the cloud, a glorious new comet may plunge into the inner Solar System.

The **Milky Way** soars overhead on these dark November nights, providing a stunning inside perspective on the huge Galaxy that is our home in space. And this month is the ideal time to view two of our nearest neighbours in the cosmos: the **Andromeda Galaxy** and its fainter cousin the **Triangulum Galaxy** (see this month's Object). Light has taken so long to travel from them that – when you gaze upon these cosmic beauties – you are looking back over 2 million years in time.

NOVEMBER'S CONSTELLATION

Perseus and neighbouring **Cassiopeia** are two of the best-loved constellations in the northern night sky. They never set as seen from Britain, and are packed with celestial goodies. In legend, Perseus was the superhero who slew Medusa, the Gorgon. Its brightest star, **Mirfak** ('elbow' in Arabic), lies 500 light years away and is 7000 times more luminous than our Sun.

But the 'star' of Perseus has to be **Algol** (whose name stems from the Arabic *al-Ghul*, meaning the 'demon star'). It represents the eye of Medusa – and it winks. Its variations were first brought to the attention of the astronomical community by 18-year-old John Goodricke, a profoundly deaf amateur astronomer. He correctly surmised that the star's strange behaviour was caused by a fainter star eclipsing a brighter one (there are actually three stars in the system). Over a period of 2.9 days, Algol dims from magnitude +2.1 to +3.4.

Another gem in Perseus – or to be exact, two of them – is the **Double Cluster**, h and chi Persei. Lying between Perseus and Cassiopeia, the duo is visible to the unaided eye, and is a sensational sight in binoculars. Some 7700 light years distant, the clusters are made of bright young blue stars. Both are a mere 12.5 million

▼ *The sky at 10 pm in mid-November, with Moon positions at three-day intervals either side of Full Moon. The star positions are also correct for 11 pm at the beginning of November, and 9 pm at the end of the month. The planets move slightly relative to the stars during the month.*

years old (compare this to our Sun, which has notched up 4.6 *billion* years so far!).

PLANETS ON VIEW

There's a dearth of bright planets in the evening sky, though with binoculars you can seek out the two most distant worlds. **Uranus** lies in Pisces: at magnitude +5.7, it sets about 4 am. In Aquarius, fainter **Neptune** (magnitude +7.9) is setting around midnight.

Serious planetary action occurs in the wee small hours. On the first few mornings of November, we're treated to the lovely tableau of brilliant Venus passing less than a degree from Mars, with Jupiter to the upper right.

Jupiter, at magnitude −1.7 in Leo, rises ever earlier: from 2 am to 0.30 am as the month progresses. In contrast, **Venus** is heading down towards the Sun. Starting the month with a rise-time of 2.30 am, by the end of November the Morning Star doesn't get out of bed until 3.30 am, when it's lying near Virgo's brightest star, Spica. Venus is brilliant at magnitude −4.2; with a small telescope, you can see it's currently half-lit by the Sun.

After Venus drops below **Mars**, the Red Planet remains between Venus and Jupiter in Virgo throughout the month, much fainter than either at magnitude +1.6, and rising at 2.30 am. **Mercury** and **Saturn** are lost in the Sun's glare this month.

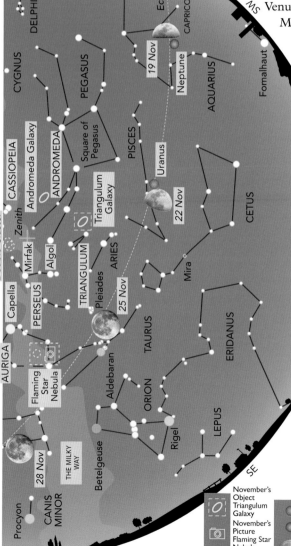

	MOON	
Date	**Time**	**Phase**
3	12.24 pm	Last Quarter
11	5.47 pm	New Moon
19	6.27 am	First Quarter
25	10.44 pm	Full Moon

November's Object Triangulum Galaxy

November's Picture Flaming Star Nebula

Radiant of Leonids

Uranus

Neptune

Moon

MOON

The star next to the Moon in the early hours of **5 November** is Regulus. On the morning of **6 November**, the Moon lies next to Jupiter; it's close to Venus and Mars on **7 November** (see Special Events); and on **8 November** the thin crescent Moon lies below all three planets.

SPECIAL EVENTS

In the morning of **7 November**, early birds are treated to a beautiful sight as the crescent Moon forms a tight triangle with Venus and Mars, with Jupiter standing higher in the sky.

The night of **17/18 November** sees the maximum of the **Leonid** meteor shower. You'll get the best views late in the evening and overnight, when the Moon has set.

NOVEMBER'S OBJECT

This month, we have a little-known extragalactic neighbour, the **Triangulum Galaxy**. To find the elusive beast, first home in on the line of stars making up the constellation of **Andromeda**. Now, locate the great **Andromeda Galaxy** – just above the line. Next – preferably with a binoculars or a low-powered telescope – move *below* the line to the same extent. With dark unpolluted skies, you should hit another fuzzy patch of light. This is the Triangulum Galaxy (M33), the third largest galaxy in the Local Group. Under exceptional conditions, it is *just* visible to the unaided eye (although it helps to be in a desert!).

To see any detail in this scruffy little spiral is a challenge even through a telescope. It has a very low surface brightness, and – at around 3 million light years distant – it is slightly further away than the Andromeda Galaxy.

▼ *Observing through an 80 mm Orion ED80T telescope, coupled to a cooled QHY8L CCD camera, Mike Wilson in Farnborough, Hampshire, captured this image of the Flaming Star Nebula (right) and IC 410/NGC 1893 (left) in Auriga. He used a light-pollution filter – the Astronomik CLS.*

👁 *Viewing tip*
The Andromeda and
Triangulum Galaxies are
often described as the
furthest objects 'easily
visible to the unaided eye'.
But they are not *that* easy
to see – especially if you
are suffering from light
pollution. The trick is
to memorize the star
patterns in Andromeda
and Triangulum, and then
to look slightly to the *side*
of where you expect each
of the galaxies to be. This
technique – called 'averted
vision' – causes the image
to fall on the outer parts
of the retina. These are
more light-sensitive than
the central region, which
has evolved to discern
the finest detail.

M33 is home to 40 billion stars: one-tenth the population of
our Milky Way. But it's making up for this lack of stars. The
Triangulum Galaxy is a hotbed of star formation, boasting an
enormous star-factory, NGC 604, which measures 1500 light
years across (as compared to the Orion Nebula's 24 light years).

This phenomenal outburst of starbirth may have been trig-
gered by an altercation between M33 and the Andromeda
Galaxy, which has also created a tenuous bridge of hydrogen
gas between the two galaxies. Their next cosmic dust-up – not
due for several billion years – may prove decisive. The two
galaxies could merge, to create a single giant elliptical galaxy....

NOVEMBER'S PICTURE

The constellation of **Auriga** (the Charioteer) is rapidly climbing into its ascendant position in our evening skies this month.
Soon, its brightest star, brilliant yellow **Capella**, will claim its
position at the zenith of the heavens for the rest of winter.

Auriga has some lesser-known gems, too, including the two
nebulae featured in this image, the **Flaming Star Nebula** and
IC 410, which surrounds the star cluster NGC 1893.

NOVEMBER'S TOPIC
General Relativity

Exactly a century ago – on 25 November 1915 – the
Prussian Academy in Berlin heard a lecture entitled 'The
Field Equations of Gravitation'. Hardly Earth-shattering,
you might think. But the lecturer was Albert Einstein, and
this talk turned the Universe on its head.

For 230 years, scientists had used Sir Isaac Newton's
theory of gravity. But it couldn't explain how Mercury's
orbit swings around the Sun, and it didn't fit with Einstein's
existing theory of Special Relativity.

The new General Relativity of 1915 agreed precisely
with Mercury's motion. It also predicted how the Sun's
gravity should bend light from distant stars – spectacularly
confirmed during a total eclipse in 1919, when stars could
be photographed close to the Sun.

Einstein explained gravity as a warping of space and time.
His theory has been tested ever more rigorously in recent
years, with precision checks on radio signals from spacecraft,
and closely orbiting pulsars (see August's Topic).

General Relativity also predicts black holes, and the expansion of the Universe from the Big Bang. Closer to home, the
Global Positioning System of satellites has to include calcu-
lations from General Relativity: if they used Newton's old
theory, your satnav would go adrift by 10 kilometres per day!

This month sees the shortest day and the longest night. And the glorious constellations of winter are riding high in the sky, in all their splendour. **Orion**, with his hunting dogs **Canis Major** and **Canis Minor**, is dominating the heavens, fighting his adversary **Taurus** (the Bull). This may be the darkest month, but we're treated to a welcome fireworks display on **13 December** as the **Geminid** meteors streak into our atmosphere.

▼ *The sky at 10 pm in mid-December, with Moon positions at three-day intervals either side of Full Moon. The star positions are also correct for 11 pm at*

DECEMBER'S CONSTELLATION

Spectacular **Orion** is one of the rare star groupings that looks like its namesake – a giant of a man with a sword below his belt, wielding a club above his head. Orion is fabled in mythology as the ultimate hunter.

The constellation contains one-tenth of the brightest stars in the sky: its seven main stars all lie in the 'top 70' brilliant stars. Despite its distinctive shape, most of these stars are not closely associated with each other – they simply line up, one behind the other.

Closest is the star that forms the hunter's right shoulder, **Bellatrix**, at 250 light years. Next is blood-red **Betelgeuse** at the top left of Orion, 640 light years away.

The brightest star in the constellation is blue-white **Rigel**, a vigorous young star more than twice as hot as our Sun, and nearly 120,000 times more luminous. Rigel lies around 860 light years from us. **Saiph**, which marks the other corner of Orion's tunic, is around 650 light years distant The two outer stars of the belt, **Alnitak** (left) and **Mintaka** (right), lie 740 and 900 light years away, respectively.

We travel 1300 light years from home to reach the middle star of the belt, **Alnilam**. And at the same distance, we see the stars of the 'sword' hanging below the belt – the lair of the great **Orion Nebula** – an enormous star-forming region estimated to measure 24 light years across.

the beginning of December, and 9 pm at the end of the month. The planets move slightly relative to the stars during the month.

PLANETS ON VIEW

In the early evening sky, there's little to offer on the planetary front until the last few days of the year, when **Mercury** appears very low in the south-west after sunset. The innermost planet is shining at magnitude -0.4 when it reaches greatest eastern elongation on **29 December**.

Uranus lies in Pisces, at magnitude $+5.8$ and setting around 2 am; while **Neptune** (magnitude $+7.9$), in Aquarius, sets at about 10 pm.

Giant planet **Jupiter** rises before midnight at month's end, dominating the sky as it blazes in Leo at magnitude -1.9.

Mars lies in Virgo, comparatively faint at magnitude $+1.4$. Rising at 2 am, it passes Spica on Christmas Eve (**24 December**) – there's a striking colour contrast between the Red Planet and the blue-white star (best seen in binoculars).

If you're up before dawn, look out for the brilliant Morning Star in the south-east: **Venus** is rising around 4 am, at an incandescent magnitude -4.0. Starting the month near Spica, it zooms through to Libra by the end of the year, when you may just glimpse **Saturn** (magnitude $+0.6$) emerging from the dawn glow to the lower left of Venus.

MOON

The star near the Moon on **1 December** is Regulus. The Moon passes under brilliant Jupiter in the wee small hours of **4 December**, and below Mars early on **6 December**. Before dawn on

WEST

AQUARIUS
PEGASUS
Ecliptic
19 Dec
Square of Pegasus
Uranus
CETUS
ANDROMEDA
TRIANGULUM
PISCES
Mira
ARIES
ALGOL
PERSEUS
22 Dec
ERIDANUS
Zenith
Capella
Pleiades
Hyades
TAURUS
AURIGA
Aldebaran
Bellatrix
Mintaka
Rigel
GEMINI
Betelgeuse
Alnilam
Alnitak
Orion Nebula
Saiph
ORION
LEPUS
COLUMBA
Radiant of Geminids
Castor
Pollux
25 Dec
Procyon
CANIS MINOR
THE MILKY WAY
Sirius
CANIS MAJOR
Adhara
CANCER
Praesepe
28 Dec
HYDRA
SOUTH
SE
EAST

| | December's Object The Hyades | | December's Picture Orion | | Radiant of Geminids |

Uranus

Moon

MOON		
Date	**Time**	**Phase**
3	7.40 am	Last Quarter
11	10.29 am	New Moon
18	3.14 pm	First Quarter
25	11.11 am	Full Moon

7 December, the crescent Moon lies near Venus, with Spica to the right; and you'll find the slender Moon below Venus on **8 December**. The Moon occults Aldebaran on **23 December** (see Special Events). On **29 December** the Moon passes below Regulus. The bright object near the Moon on **30 and 31 December** is giant planet Jupiter.

SPECIAL EVENTS

You can see meteors from the annual **Geminid** shower on the nights of both **13/14 and 14/15 December**: with the Moon well out of the way, it's an ideal year for enjoying this abundant display of slow bright meteors, shed by the asteroid Phaethon.

The Winter Solstice occurs at 4.48 am on **22 December**. As a result of the tilt of Earth's axis, the Sun reaches its lowest point in the heavens as seen from the northern hemisphere: we get the shortest days, and the longest nights.

On **23 December**, the Moon occults Aldebaran. Depending on where you are in Britain, the star disappears some time between 6.05 pm and 6.25 pm, and reappears between 7.05 pm and 7.20 pm.

DECEMBER'S OBJECT

The 'V'-shaped **Hyades** star cluster, which forms the 'head' of **Taurus** (the Bull), doesn't hold a candle to the dazzling **Pleiades**. But it's the nearest star cluster to the Earth, and it forms the first rung of the ladder in establishing the cosmic distance scale.

In legend, the Hyades often feature as female figures, including the nymphs who cared for Bacchus as a baby. But we prefer the Roman interpretation as 'little pigs'; or the Chinese vision of a 'rabbit net'.

◄ In Oxfordshire, Robin Scagell took this classic photograph of Orion. It was captured on Ektachrome 400 film, with a Canon A-1 camera, exposed for 3 minutes. He used a 55 mm lens at f/2.8, and added a diffraction filter to show the stars more sharply.

Scagell points out that – if you want to bring out the ruddiness of the Orion Nebula – using film is better at highlighting the colour, as digital photography is not as red-sensitive.

Although **Aldebaran**, marking the bull's angry eye, looks as though it's part of the Hyades, this red giant just happens to lie in the same direction, and at less than half the distance. The Hyades cluster lies 153 light years away, and contains about 200 stars, all around 625 million years old – very young on the stellar scale. They could have a celestial twin, too. **Praesepe** – the Beehive Cluster in **Cancer** – is the same age, and its stars are moving in the same direction. It may well be that the two clusters share a common birth.

DECEMBER'S PICTURE

Icon of the winter skies, the constellation of **Orion** is climbing to take over centre-stage in the heavens. It's a picture-gallery of brilliant star colours. Baleful red giant Betelgeuse hogs the attention as Orion's left shoulder, while blue-white Rigel marks the hunter's right ankle. The other stars are all surpassed by a marvel of starbirth, the glorious Orion Nebula. This fuzzy red blur lies in Orion's sword, dangling below his belt of three stars.

DECEMBER'S TOPIC
Star names

Why do the brightest stars have such strange names? The reason is that they date from antiquity, and have been passed on down generations ever since. The original western star names – like the original constellations – were probably Babylonian or Chaldean, but few of these survive. The Greeks took up the baton after that, and the name of the star Antares in Scorpius (see July's Constellation) is a direct result. It means 'rival of Ares' because its red colour rivals that of the planet Mars (*Ares* in Greek).

But the Arabs were largely responsible for the star names we have inherited today. Working in the so-called 'Dark Ages' between the 6th and 10th centuries AD, they took over the naming of the sky – hence the number of stars beginning with the letters 'al' (Arabic for 'the'). **Algol**, in the constellation **Perseus**, means 'the demon' – possibly because the Arabs noticed that its brightness seems to 'wink' every few days. **Deneb**, in **Cygnus**, also has Arabic roots – it means 'the tail' (of the flying bird).

The most famous star in the sky has to be **Betelgeuse** – known to generations of school kids as 'Beetlejuice'. It was gloriously interpreted to mean 'the armpit of the central one'. But the 'B' in Betelgeuse turned out to be a mistransliteration – and so we're none the wiser as to how our distant ancestors really identified this fiery red star.

There's always something to see in our Solar System, from planets to meteors or the Moon. These objects are very close to us – in astronomical terms – so their positions, shapes and sizes appear to change constantly. It is important to know when, where and how to look if you are to enjoy exploring Earth's neighbourhood. Here we give the best dates in 2015 for observing the planets and meteors (weather permitting!), and explain some of the concepts that will help you to get the most out of your observing.

THE INFERIOR PLANETS

A planet with an orbit that lies closer to the Sun than the orbit of Earth is known as *inferior*. Mercury and Venus are the inferior planets. They show a full range of phases (like the Moon) from the thinnest crescents to full, depending on their position in relation to the Earth and the Sun. The diagram below shows the various positions of the inferior planets. They are invisible when at *conjunction*, when they are either behind the Sun, or between the Earth and the Sun, and lost in the latter's glare.

Magnitudes
Astronomers measure the brightness of stars, planets and other celestial objects using a scale of *magnitudes*. Somewhat confusingly, fainter objects have higher magnitudes, while brighter objects have lower magnitudes; the most brilliant stars have negative magnitudes! Naked-eye stars range from magnitude −1.5 for the brightest star, Sirius, to +6.5 for the faintest stars you can see on a really dark night.
  As a guide, here are the magnitudes of selected objects:

Sun	−26.7
Full Moon	−12.5
Venus (at its brightest)	−4.7
Sirius	−1.5
Betelgeuse	+0.4
Polaris (Pole Star)	+2.0
Faintest star visible to the naked eye	+6.5
Faintest star visible to the Hubble Space Telescope	+31

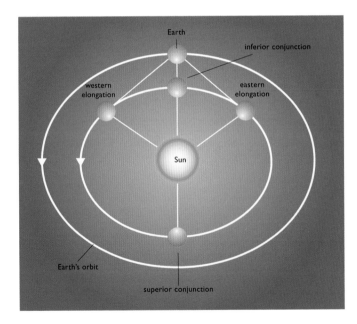

◀ At eastern or western elongation, an inferior planet is at its maximum angular distance from the Sun. Conjunction occurs at two stages in the planet's orbit. Under certain circumstances, an inferior planet can transit across the Sun's disc at inferior conjunction.

Mercury

Mercury has two good evening appearances: in January and again in May. The August–September evening apparition is completely lost in the bright twilight sky, but the planet is visible again at the end of December. You can see Mercury in the morning skies of February, though dawn twilight will wash out its June–July apparition. The innermost planet reserves its best morning display for October.

⬤ Maximum elongations of Mercury in 2015	
Date	Separation
14 January	19° east
24 February	27° west
7 May	21° east
24 June	23° west
4 September	27° east
16 October	18° west
29 December	20° east

Maximum elongation of Venus in 2015	
Date	Separation
6 June	45° east
26 October	46° west

Venus

As the Evening Star, Venus dominates the sky after nightfall for the first half of the year. After swinging between the Earth and the Sun on 15 August, Venus reappears – equally resplendent – as the Morning Star through to the end of 2015.

THE SUPERIOR PLANETS

The superior planets are those with orbits that lie beyond that of the Earth. They are Mars, Jupiter, Saturn, Uranus and Neptune. The best time to observe a superior planet is when the Earth lies between it and the Sun. At this point in a planet's orbit, it is said to be at *opposition*.

▶ Superior planets are invisible at conjunction. At quadrature the planet is at right angles to the Sun as viewed from Earth. Opposition is the best time to observe a superior planet.

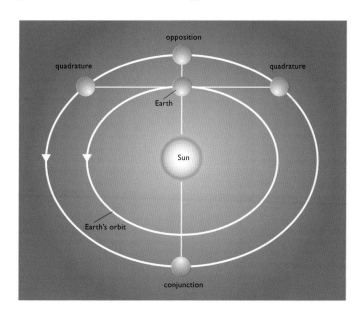

Progress of Mars through the constellations	
Early January	Capricornus
Mid Jan–mid February	Aquarius
Late February–March	Pisces
April	Aries
August	Cancer
September–October	Leo
November–December	Virgo

Mars

The Red Planet spends most of the year skulking dimly near the horizon. It's in the evening sky until April, and then reappears in the dawn twilight.

Jupiter

The giant planet starts the year in Leo, and moves into Cancer for its opposition on 6 February, when it's visible all night long. After a spectacular meeting with Venus in late July, Jupiter drops into the twilight glow. It reappears in the morning sky late in September, back in Leo, where it resides for the rest of the year.

Saturn

You'll find the ringed planet on the border of Libra and Scorpius throughout the year. Saturn is prominent in the morning sky at the start of 2015; it is visible all night when it reaches opposition on 23 May, and remains a feature of the evening sky until October.

Uranus

Just perceptible to the naked eye, Uranus is visible in the evening sky from January to March, and then from June to December. It swims among the stars of Pisces all year, and is at opposition on 12 October.

Neptune

Lying in Aquarius all year, the most distant planet is at opposition on 1 September. Neptune can be seen – though only through binoculars or a telescope – in January and February, and then from May to the end of the year.

SOLAR AND LUNAR ECLIPSES

Solar Eclipses

The two solar eclipses in 2015 take it in turns to favour the North and South Poles. On 20 March, a total eclipse of the Sun is visible from parts of the Arctic Ocean, including the Faroe Islands and Svalbard. It appears as a partial eclipse from north-west Asia, northern Africa and all of Europe including the British Isles, where up to 98% of the Sun will be covered (see March's Special Events). The partial solar eclipse of 13 September can be seen from southern Africa, the southern Indian Ocean and parts of Antarctica.

Lunar Eclipses

The year 2015 has two total lunar eclipses, visible from more-or-less opposite sides of the Earth. The total eclipse of 4 April can be seen from eastern Asia, Australasia, the Pacific Ocean and western parts of North America: it's not visible from

Astronomical distances

For objects in the Solar System, such as the planets, we can give their distances from the Earth in kilometres. But the distances are just too huge once we reach out to the stars. Even the nearest star (Proxima Centauri) lies 25 million million kilometres away.

So astronomers use a larger unit – the *light year*. This is the distance that light travels in one year, and it equals 9.46 million million kilometres.

Here are the distances to some familiar astronomical objects, in light years:

Proxima Centauri	4.2
Betelgeuse	640
Centre of the Milky Way	27,000
Andromeda Galaxy	2.5 million
Most distant galaxies seen by the Hubble Space Telescope	13 billion

◄ Where the dark central part (the umbra) of the Moon's shadow reaches the Earth, we see a total eclipse. People located within the penumbra see a partial eclipse. If the umbral shadow does not reach the Earth, we see an annular eclipse. This type of eclipse occurs when the Moon is at a distant point in its orbit and is not quite large enough to cover the whole of the Sun's disc.

Britain. On 28 September, we're treated to a total lunar eclipse that's visible from western Asia, Africa, the Pacific Ocean, the Americas and Europe – including the British Isles.

METEOR SHOWERS

Shooting stars – or *meteors* – are tiny particles of interplanetary dust, known as *meteoroids*, burning up in the Earth's atmosphere. At certain times of year, the Earth passes through a stream of these meteoroids (usually debris left behind by a comet) and we see a *meteor shower*. The point in the sky from which the

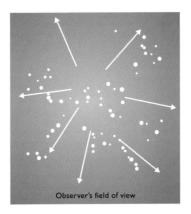

Observer's field of view

▶ Meteors from a common source, occurring during a shower, enter the atmosphere along parallel trajectories. As a result of perspective, however, they appear to diverge from a single point in the sky – the radiant.

meteors appear to emanate is known as the *radiant*. Most showers are known by the constellation in which the radiant is situated.

When watching meteors for a co-ordinated meteor programme, observers generally note the time, seeing conditions, cloud cover, their own location, the time and brightness of each meteor, and whether it was from the main meteor stream. It is also worth noting details of persistent afterglows (trains) and fireballs, and making counts of how many meteors appear in a given period.

COMETS

Comets are small bodies in orbit about the Sun. Consisting of frozen gases and dust, they are often known as 'dirty snowballs'. When their orbits bring them close to the Sun, the ices evaporate and dramatic tails of gas and dust can sometimes be seen.

A number of comets move round the Sun in fairly small, elliptical orbits in periods of a few years; others have much longer periods. Most really brilliant comets have orbital periods of several thousands or even millions of years. The exception is Comet Halley, a bright comet with a period of about 76 years. It was last seen with the naked eye in 1986.

Binoculars and wide-field telescopes provide the best views of comet tails. Larger telescopes with a high magnification are necessary to observe fine detail in the gaseous head (*coma*). Most comets are discovered with professional instruments, but a few are still found by experienced amateur astronomers.

No known comets will reach naked-eye brightness this year, but a brilliant new comet could always put in a surprise appearance.

Angular separations

Astronomers measure the distance between objects, as we see them in the sky, by the angle between the objects in degrees (symbol °). From the horizon to the point above your head is 90 degrees. All around the horizon is 360 degrees.

You can use your hand, held at arm's length, as a rough guide to angular distances, as follows:

Width of index finger 1°
Width of clenched hand 10°
Thumb to little finger
  on outspread hand 20°

For smaller distances, astronomers divide the degree into 60 arc minutes (symbol ′), and the arc minute into 60 arc seconds (symbol ″).

Deep-sky objects are 'fuzzy patches' that lie outside the Solar System. They include star clusters, nebulae and galaxies. To observe the majority of deep-sky objects you will need binoculars or a telescope, but there are also some beautiful naked-eye objects, notably the Pleiades and the Orion Nebula.

The faintest object that an instrument can see is its *limiting magnitude*. The table gives a rough guide, for good seeing conditions, for a variety of small- to medium-sized telescopes.

We have provided a selection of recommended deep-sky targets, together with their magnitudes. Some are described in more detail in our monthly 'Object' features. Look on the appropriate month's map to find which constellations are on view, and then choose your objects using the list below. We have provided celestial coordinates for readers with detailed star maps or Go To telescopes. The suggested times of year for viewing are when the constellation is highest in the sky in the late evening.

Limiting magnitude for small to medium telescopes	
Aperture (mm)	Limiting magnitude
50	+11.2
60	+11.6
70	+11.9
80	+12.2
100	+12.7
125	+13.2
150	+13.6

RECOMMENDED DEEP-SKY OBJECTS

Andromeda – autumn and early winter

M31 (NGC 224) Andromeda Galaxy	3rd-magnitude spiral galaxy RA 00h 42.7m Dec +41° 16'
M32 (NGC 221)	8th-magnitude elliptical galaxy, a companion to M31 RA 00h 42.7m Dec +40° 52'
M110 (NGC 205)	8th-magnitude elliptical galaxy RA 00h 40.4m Dec +41° 41'
NGC 7662 Blue Snowball	8th-magnitude planetary nebula RA 23h 25.9m Dec +42° 33'

Aquarius – late autumn and early winter

M2 (NGC 7089)	6th-magnitude globular cluster RA 21h 33.5m Dec –00° 49'
M72 (NGC 6981)	9th-magnitude globular cluster RA 20h 53.5m Dec –12° 32'
NGC 7293 Helix Nebula	7th-magnitude planetary nebula RA 22h 29.6m Dec –20° 48'
NGC 7009 Saturn Nebula	8th-magnitude planetary nebula RA 21h 04.2m Dec –11° 22'

Aries – early winter

NGC 772	10th-magnitude spiral galaxy RA 01h 59.3m Dec +19° 01'

Auriga – winter

M36 (NGC 1960)	6th-magnitude open cluster RA 05h 36.1m Dec +34° 08'
M37 (NGC 2099)	6th-magnitude open cluster RA 05h 52.4m Dec +32° 33'
M38 (NGC 1912)	6th-magnitude open cluster RA 05h 28.7m Dec +35° 50'

Cancer – late winter to early spring

M44 (NGC 2632) Praesepe or Beehive	3rd-magnitude open cluster RA 08h 40.1m Dec +19° 59'
M67 (NGC 2682)	7th-magnitude open cluster RA 08h 50.4m Dec +11° 49'

Canes Venatici – visible all year

M3 (NGC 5272)	6th-magnitude globular cluster RA 13h 42.2m Dec +28° 23'
M51 (NGC 5194/5) Whirlpool Galaxy	8th-magnitude spiral galaxy RA 13h 29.9m Dec +47° 12'
M63 (NGC 5055)	9th-magnitude spiral galaxy RA 13h 15.8m Dec +42° 02'
M94 (NGC 4736)	8th-magnitude spiral galaxy RA 12h 50.9m Dec +41° 07'
M106 (NGC4258)	8th-magnitude spiral galaxy RA 12h 19.0m Dec +47° 18'

Canis Major – late winter

M41 (NGC 2287)	4th-magnitude open cluster RA 06h 47.0m Dec –20° 44'

Capricornus – late summer and early autumn

M30 (NGC 7099)	7th-magnitude globular cluster RA 21h 40.4m Dec –23° 11'

Cassiopeia – visible all year

M52 (NGC 7654)	6th-magnitude open cluster RA 23h 24.2m Dec +61° 35'
M103 (NGC 581)	7th-magnitude open cluster RA 01h 33.2m Dec +60° 42'
NGC 225	7th-magnitude open cluster RA 00h 43.4m Dec +61 47'
NGC 457	6th-magnitude open cluster RA 01h 19.1m Dec +58° 20'
NGC 663	Good binocular open cluster RA 01h 46.0m Dec +61° 15'

Cepheus – visible all year

Delta Cephei	Variable star, varying between +3.5 and +4.4 with a period of 5.37 days. It has a magnitude +6.3 companion and they make an attractive pair for small telescopes or binoculars.

Cetus – late autumn

Mira (omicron Ceti)	Irregular variable star with a period of roughly 330 days and a range between +2.0 and +10.1.
M77 (NGC 1068)	9th-magnitude spiral galaxy RA 02h 42.7m Dec –00° 01'

Coma Berenices – spring

M53 (NGC 5024)	8th-magnitude globular cluster RA 13h 12.9m Dec +18° 10'
M64 (NGC 4286) Black Eye Galaxy	8th-magnitude spiral galaxy with a prominent dust lane that is visible in larger telescopes. RA 12h 56.7m Dec +21° 41'
M85 (NGC 4382)	9th-magnitude elliptical galaxy RA 12h 25.4m Dec +18° 11'
M88 (NGC 4501)	10th-magnitude spiral galaxy RA 12h 32.0m Dec.+14° 25'
M91 (NGC 4548)	10th-magnitude spiral galaxy RA 12h 35.4m Dec +14° 30'
M98 (NGC 4192)	10th-magnitude spiral galaxy RA 12h 13.8m Dec +14° 54'
M99 (NGC 4254)	10th-magnitude spiral galaxy RA 12h 18.8m Dec +14° 25'
M100 (NGC 4321)	9th-magnitude spiral galaxy RA 12h 22.9m Dec +15° 49'
NGC 4565	10th-magnitude spiral galaxy RA 12h 36.3m Dec +25° 59'

Cygnus – late summer and autumn

Cygnus Rift	Dark cloud just south of Deneb that appears to split the Milky Way in two.
NGC 7000 North America Nebula	A bright nebula against the back- ground of the Milky Way, visible with binoculars under dark skies. RA 20h 58.8m Dec +44° 20'
NGC 6992 Veil Nebula (part)	Supernova remnant, visible with binoculars under dark skies. RA 20h 56.8m Dec +31 28'
M29 (NGC 6913)	7th-magnitude open cluster RA 20h 23.9m Dec +36° 32'
M39 (NGC 7092)	Large 5th-magnitude open cluster RA 21h 32.2m Dec +48° 26'
NGC 6826 Blinking Planetary	9th-magnitude planetary nebula RA 19 44.8m Dec +50° 31'

Delphinus – late summer

NGC 6934	9th-magnitude globular cluster RA 20h 34.2m Dec +07° 24'

Draco – midsummer

NGC 6543	9th-magnitude planetary nebula RA 17h 58.6m Dec +66° 38'

Gemini – winter

M35 (NGC 2168)	5th-magnitude open cluster RA 06h 08.9m Dec +24° 20'
NGC 2392 Eskimo Nebula	8–10th-magnitude planetary nebula RA 07h 29.2m Dec +20° 55'

Hercules – early summer

M13 (NGC 6205)	6th-magnitude globular cluster RA 16h 41.7m Dec +36° 28'
M92 (NGC 6341)	6th-magnitude globular cluster RA 17h 17.1m Dec +43° 08'
NGC 6210	9th-magnitude planetary nebula RA 16h 44.5m Dec +23 49'

Hydra – early spring

M48 (NGC 2548)	6th-magnitude open cluster RA 08h 13.8m Dec –05° 48'
M68 (NGC 4590)	8th-magnitude globular cluster RA 12h 39.5m Dec –26° 45'

M83 (NGC 5236)	8th-magnitude spiral galaxy RA 13h 37.0m Dec –29° 52'
NGC 3242 Ghost of Jupiter	9th-magnitude planetary nebula RA 10h 24.8m Dec –18° 38'

Leo – spring

M65 (NGC 3623)	9th-magnitude spiral galaxy RA 11h 18.9m Dec +13° 05'
M66 (NGC 3627)	9th-magnitude spiral galaxy RA 11h 20.2m Dec +12° 59'
M95 (NGC 3351)	10th-magnitude spiral galaxy RA 10h 44.0m Dec +11° 42'
M96 (NGC 3368)	9th-magnitude spiral galaxy RA 10h 46.8m Dec +11° 49'
M105 (NGC 3379)	9th-magnitude elliptical galaxy RA 10h 47.8m Dec +12° 35'

Lepus – winter

M79 (NGC 1904)	8th-magnitude globular cluster RA 05h 24.5m Dec –24° 33'

Lyra – spring

M56 (NGC 6779)	8th-magnitude globular cluster RA 19h 16.6m Dec +30° 11'
M57 (NGC 6720) Ring Nebula	9th-magnitude planetary nebula RA 18h 53.6m Dec +33° 02'

Monoceros – winter

M50 (NGC 2323)	6th-magnitude open cluster RA 07h 03.2m Dec –08° 20'
NGC 2244	Open cluster surrounded by the faint Rosette Nebula, NGC 2237. Visible in binoculars. RA 06h 32.4m Dec +04° 52'

Ophiuchus – summer

M9 (NGC 6333)	8th-magnitude globular cluster RA 17h 19.2m Dec –18° 31'
M10 (NGC 6254)	7th-magnitude globular cluster RA 16h 57.1m Dec –04° 06'
M12 (NCG 6218)	7th-magnitude globular cluster RA 16h 47.2m Dec –01° 57'
M14 (NGC 6402)	8th-magnitude globular cluster RA 17h 37.6m Dec –03° 15'
M19 (NGC 6273)	7th-magnitude globular cluster RA 17h 02.6m Dec –26° 16'
M62 (NGC 6266)	7th-magnitude globular cluster RA 17h 01.2m Dec –30° 07'
M107 (NGC 6171)	8th-magnitude globular cluster RA 16h 32.5m Dec –13° 03'

Orion – winter

M42 (NGC 1976) Orion Nebula	4th-magnitude nebula RA 05h 35.4m Dec –05° 27'
M43 (NGC 1982)	5th-magnitude nebula RA 05h 35.6m Dec –05° 16'
M78 (NGC 2068)	8th-magnitude nebula RA 05h 46.7m Dec +00° 03'

Pegasus – autumn

M15 (NGC 7078)	6th-magnitude globular cluster RA 21h 30.0m Dec +12° 10'

Perseus – autumn to winter

M34 (NGC 1039)	5th-magnitude open cluster RA 02h 42.0m Dec +42° 47'
M76 (NGC 650/1) Little Dumbbell	11th-magnitude planetary nebula RA 01h 42.4m Dec +51° 34'

NGC 869/884 Double Cluster	Pair of open star clusters RA 02h 19.0m Dec +57° 09' RA 02h 22.4m Dec +57° 07'

Pisces – autumn

M74 (NGC 628)	9th-magnitude spiral galaxy RA 01h 36.7m Dec +15° 47'

Puppis – late winter

M46 (NGC 2437)	6th-magnitude open cluster RA 07h 41.8m Dec –14° 49'
M47 (NGC 2422)	4th-magnitude open cluster RA 07h 36.6m Dec –14° 30'
M93 (NGC 2447)	6th-magnitude open cluster RA 07h 44.6m Dec –23° 52'

Sagitta – late summer

M71 (NGC 6838)	8th-magnitude globular cluster RA 19h 53.8m Dec +18° 47'

Sagittarius – summer

M8 (NGC 6523) Lagoon Nebula	6th-magnitude nebula RA 18h 03.8m Dec –24° 23'
M17 (NGC 6618) Omega Nebula	6th-magnitude nebula RA 18h 20.8m Dec –16° 11'
M18 (NGC 6613)	7th-magnitude open cluster RA 18h 19.9m Dec –17 08'
M20 (NGC 6514) Trifid Nebula	9th-magnitude nebula RA 18h 02.3m Dec –23° 02'
M21 (NGC 6531)	6th-magnitude open cluster RA 18h 04.6m Dec –22° 30'
M22 (NGC 6656)	5th-magnitude globular cluster RA 18h 36.4m Dec –23° 54'
M23 (NGC 6494)	5th-magnitude open cluster RA 17h 56.8m Dec –19° 01'
M24 (NGC 6603)	5th-magnitude open cluster RA 18h 16.9m Dec –18° 29'
M25 (IC 4725)	5th-magnitude open cluster RA 18h 31.6m Dec –19° 15'
M28 (NGC 6626)	7th-magnitude globular cluster RA 18h 24.5m Dec –24° 52'
M54 (NGC 6715)	8th-magnitude globular cluster RA 18h 55.1m Dec –30° 29'
M55 (NGC 6809)	7th-magnitude globular cluster RA 19h 40.0m Dec –30° 58'
M69 (NGC 6637)	8th-magnitude globular cluster RA 18h 31.4m Dec –32° 21'
M70 (NGC 6681)	8th-magnitude globular cluster RA 18h 43.2m Dec –32° 18'
M75 (NGC 6864)	9th-magnitude globular cluster RA 20h 06.1m Dec –21° 55'

Scorpius (northern part) – midsummer

M4 (NGC 6121)	6th-magnitude globular cluster RA 16h 23.6m Dec –26° 32'
M7 (NGC 6475)	3rd-magnitude open cluster RA 17h 53.9m Dec –34° 49'
M80 (NGC 6093)	7th-magnitude globular cluster RA 16h 17.0m Dec –22° 59'

Scutum – mid to late summer

M11 (NGC 6705) Wild Duck Cluster	6th-magnitude open cluster RA 18h 51.1m Dec –06° 16'

M26 (NGC 6694)	8th-magnitude open cluster RA 18h 45.2m Dec –09° 24'

Serpens – summer

M5 (NGC 5904)	6th-magnitude globular cluster RA 15h 18.6m Dec +02° 05'
M16 (NGC 6611)	6th-magnitude open cluster, surrounded by the Eagle Nebula. RA 18h 18.8m Dec –13° 47'

Taurus – winter

M1 (NGC 1952) Crab Nebula	8th-magnitude supernova remnant RA 05h 34.5m Dec +22° 00'
M45 Pleiades	1st-magnitude open cluster, an excellent binocular object. RA 03h 47.0m Dec +24° 07'

Triangulum – autumn

M33 (NGC 598)	6th-magnitude spiral galaxy RA 01h 33.9m Dec +30° 39'

Ursa Major – all year

M81 (NGC 3031)	7th-magnitude spiral galaxy RA 09h 55.6m Dec +69° 04'
M82 (NGC 3034)	8th-magnitude starburst galaxy RA 09h 55.8m Dec +69° 41'
M97 (NGC 3587) Owl Nebula	12th-magnitude planetary nebula RA 11h 14.8m Dec +55° 01'
M101 (NGC 5457)	8th-magnitude spiral galaxy RA 14h 03.2m Dec +54° 21'
M108 (NGC 3556)	10th-magnitude spiral galaxy RA 11h 11.5m Dec +55° 40'
M109 (NGC 3992)	10th-magnitude spiral galaxy RA 11h 57.6m Dec +53° 23'

Virgo – spring

M49 (NGC 4472)	8th-magnitude elliptical galaxy RA 12h 29.8m Dec +08° 00'
M58 (NGC 4579)	10th-magnitude spiral galaxy RA 12h 37.7m Dec +11° 49'
M59 (NGC 4621)	10th-magnitude elliptical galaxy RA 12h 42.0m Dec +11° 39'
M60 (NGC 4649)	9th-magnitude elliptical galaxy RA 12h 43.7m Dec +11° 33'
M61 (NGC 4303)	10th-magnitude spiral galaxy RA 12h 21.9m Dec +04° 28'
M84 (NGC 4374)	9th-magnitude elliptical galaxy RA 12h 25.1m Dec +12° 53'
M86 (NGC 4406)	9th-magnitude elliptical galaxy RA 12h 26.2m Dec +12° 57'
M87 (NGC 4486)	9th-magnitude elliptical galaxy RA 12h 30.8m Dec +12° 24'
M89 (NGC 4552)	10th-magnitude elliptical galaxy RA 12h 35.7m Dec +12° 33'
M90 (NGC 4569)	9th-magnitude spiral galaxy RA 12h 36.8m Dec +13° 10'
M104 (NGC 4594) Sombrero Galaxy	Almost edge-on 8th-magnitude spiral galaxy. RA 12h 40.0m Dec –11° 37'

Vulpecula – late summer and autumn

M27 (NGC 6853) Dumbbell Nebula	8th-magnitude planetary nebula RA 19h 59.6m Dec +22° 43'

DEALING WITH LIGHT POLLUTION

Few of us these days are free from light pollution. The majority of us live either in towns or cities, or not far from one. And even in the countryside, miles from a major town, the skies still betray the presence of urban areas by the orange glow that they cast. So can technology come to the rescue?

Let's make one thing clear to start with. There is no 'magic filter' that we can hold up to a light-polluted sky that will allow the stars and other objects to shine through. But there are some filters that can help to a greater or lesser extent, and there are also ways of maximizing our stargazing experience. And of course observers of the Sun, Moon and planets have no trouble at all with light pollution, so it is just with what people call the 'faint fuzzies' that the problem arises.

UNDERSTANDING THE PROBLEM

When we see a light-polluted sky, we tend to blame just the streetlights. But they are only part of the problem. True, the sky is generally the same yellow-orange colour of streetlights, but what are they shining on? Much of it is the water vapour that results from the British Isles being an island in the path of westerly winds from the Atlantic. The more vapour, the more milky our skies by day and the worse the light pollution by night as the streetlights shine on the general haze. So the simplest way of observing those faint objects from your own location is to wait for the nights when it really is clear. These usually occur when we have cold, polar air washing over the country.

But regrettably, these good conditions apply on only a few nights of the year, usually those on which we have some other commitment that we can't wriggle out of. So on run-of-the-mill clear nights which are more frequent, we may turn to filters to help us.

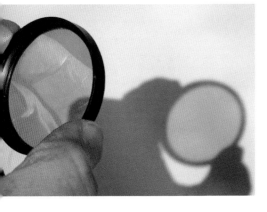

▼ *A typical light-pollution filter, here an Astronomik CLS filter in 50 mm size, is an interference filter. Its multilayer coatings transmit some colours – here, blue – while reflecting others, such as magenta and green, depending on the angle at which the light passes through it.*

THE CHANGING COLOUR OF STREETLIGHTS

Most streetlights used in the UK are either the older orange low-pressure sodium, still found on many main and suburban roads, and the more recent pinkish high-pressure sodium. LED lights are now taking over from both on account of their white light and longer lifetime. A few of the blue-white mercury lights are still to be found, but as from April this year their sale is banned in the EU.

Over the years, our night skies have changed colour from the vivid orange of low-pressure sodium, which has only a single wavelength, to the much broader spectrum of the more modern streetlights.

In addition, around town and city centres there are other light sources from advertising signs and floodlights which add to the whiteness of the sky colour.

DEEP-SKY FILTERS

Many sets of eyepieces come with an array of coloured filters that screw into the eyepiece barrel. These are intended for planetary use, and are occasionally useful for improving the contrast of markings on planets. However, they have no effect on light pollution. Instead, we have to turn to more selective filters that cut out only restricted parts of the spectrum. The reason that filters work at all is in cases where the objects that we want to observe and the light sources in the light pollution occupy different parts of the spectrum.

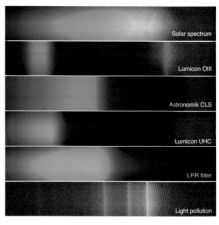

▲ *How various light-pollution filters absorb different parts of the spectrum. At bottom is the spectrum of a typical UK sky, with a strong yellow line from low-pressure sodium and other lines from high-pressure sodium.*

Most nebulae – shining gas clouds – emit fairly well-defined colours of light, depending on the gases within them. A bright nebula such as the Orion Nebula contains mostly hydrogen, which has a very specific spectrum consisting of particular wavelengths of deep red, blue-green and violet light. Planetary nebulae also shine at particular wavelengths, with some differences due to stronger nitrogen and oxygen emissions. So there is the possibility of using a filter that will transmit the light from nebulae while rejecting that from the yellowish streetlights. In practice, there is quite a bit of overlap, but there are still enough differences to be useful.

Galaxies, on the other hand, emit more or less white light, consisting of all colours. So there is no possibility of isolating just the light from a galaxy, though we can use filters that cut down the colours where streetlights predominate to give a little assistance. However, the latest breed of streetlights, using white LEDs, have virtually the same colour output as galaxies, so as these lights increase in use, filters will become less useful.

Filters break down into broadband and narrowband filters. They fit into the eyepiece of your telescope using a standard screw thread, in either 31.7 mm or 50 mm size, but are not easily adapted for use with binoculars or non-standard instruments. The broadband variety cut out the yellowish chunks of the spectrum where the streetlights dominate, an example being the Lumicon Deep Sky Filter. They are often generically known as LPR (Light Pollution Rejection) filters. Then there are narrowband filters, which allow through only very specific wavelengths, notably the light from the oxygen in nebulae, known as O III (O-three) filters. Another popular type, the UHC (Ultra High Contrast) filter, allows through O III and also some other adjacent wavelengths from nitrogen.

As these transmit more light than O III filters, they give a brighter view. Typically, a 31.7 mm filter will cost from £60 to £100, with 50 mm filters costing up to £180, though there are budget filters of all types available.

The usefulness of filters depends very much on your locality, your local streetlighting and the objects that you want to view. Many filters were developed and tested some years ago in the US, which then had a much higher usage of mercury lighting than in the UK, though this has now changed. So manufacturers' claims may not apply to your own local circumstances. Filters in general work better with nebulae rather than galaxies or star clusters. And each nebula has a slightly different spectrum, depending on the gases present and the level of excitation of those gases, so what improves the view of one nebula may make the view of another worse.

If you are faced with really bad light pollution near a city centre, probably not even the costliest filter used with a large telescope will have much effect. The effect, if any, will be marginal.

In the suburbs, particularly where there is still a lot of low-pressure sodium lighting, broadband deep-sky filters can help to dim the yellow lights and improve your views of nebulae and to a much lesser extent galaxies, though the results are never spectacular. And oddly enough, they help most in country areas, where there is only a fairly small amount of light pollution to deal with. Even in what seem to be pitch-black skies, broadband filters can improve the view of nebulae.

The most dramatic results come with the use of the more narrowband filters to view nebulae in areas of moderate to low light pollution. An object such as the Veil Nebula in Cygnus is invisible from many locations, but it can become easily seen using a UHC or O III filter. The latter in particular is more use with telescopes over about 130 mm aperture, as it does darken

▶ The light from a planetary nebula such as M27 in Vulpecula consists of specific wavelengths of green and red light from oxygen and hydrogen, so deep-sky filters improve its visibility. This photo was taken from light-polluted Newport, south Wales, by Nick Hart.

the field of view considerably. This is the nearest to a magic filter that exists, but it only works well on certain objects.

By and large, these filters tend to work only with telescopes or sometimes binoculars. Just holding them up to the night sky and hoping to catch a glimpse of deep-sky objects is not usually successful, because the objects on view are quite small and we need optical power to make them large enough to be easily visible.

GETTING AWAY FROM IT ALL

Technology notwithstanding, there's no substitute for a dark sky. But where's the best place to go? To get the best dark skies in the UK, you really need to head for those places that are as remote as possible from large conurbations. Certain areas are now recognized as Dark Sky places – Galloway Forest Dark Sky Park, Sark Dark Sky Island in the Channel Islands, Exmoor Dark Sky Reserve and Brecon Beacons National Park. These offer a good combination of dark skies and public access. But are they the darkest places you can find in the British Isles? In each case, they are not very far from built-up areas, and by no means are they as dark as truly remote spots. Even a small nearby town creates its own bubble of light pollution.

The far north-west Highlands and Hebrides are farther from city lights than any other place in Britain. Certain parts of mid Wales are also particularly dark, while many areas in the west of Ireland are also well away from cities. But one further requirement is a good chance of clear skies, and finding both dark and clear skies can be a challenge. There are no good statistics on night-time clear skies nationwide, though sunshine statistics could be good enough. And if you are fortunate, you could be rewarded with amazing views of the night sky.

◀ *The Milky Way in Sagittarius photographed from one of the darkest skies in England, in west Cornwall.*